Praise for

fast
asleep

"A brilliantly informative book showing you how to boost your sleep and cure insomnia."

—*Daily Mail* (London)

"Mosley's firsthand experience of insomnia, and his search for answers, makes this book a real page-turner. It will change many lives."

—Prof. Barry Marshall, winner of the
Nobel Prize in Physiology or Medicine

"I read this eagerly because I am desperate for tips on how to sleep better. It is based around the revolutionary idea that when it comes to sleep, what matters is not the hours you spend in bed but the quality of the sleep you are getting—your sleep efficiency. This book was full of surprises!"

—Jeremy Vine, BBC Radio 2 presenter
and broadcast journalist

"A fascinating and accessible book with some truly surprising findings. If you or someone you know struggles with poor sleep, this book is a must."

—Prof. Felice Jacka, director of the
Food & Mood Centre at Deakin University

Also by Dr. Michael Mosley

The FastDiet

FastExercise

The FastLife

The 8-Week Blood Sugar Diet

The Clever Gut Diet

The Fast800 Diet

fast asleep

Improve Brain Function,
Lose Weight, Boost Your Mood,
Reduce Stress, and Become
a Better Sleeper

DR. MICHAEL MOSLEY

ATRIA PAPERBACK
New York • London • Toronto • Sydney • New Delhi

ATRIA
PAPERBACK

An Imprint of Simon & Schuster, Inc.
1230 Avenue of the Americas
New York, NY 10020

Originally published in Great Britain in 2020 by Short Books

First Atria Paperback edition July 2021

ATRIA PAPERBACK and colophon are trademarks of Simon & Schuster, Inc.

For information about special discounts for bulk purchases, please contact Simon & Schuster Special Sales at 1-866-506-1949 or business@simonandschuster.com.

The Simon & Schuster Speakers Bureau can bring authors to your live event. For more information or to book an event, contact the Simon & Schuster Speakers Bureau at 1-866-248-3049 or visit our website at www.simonspeakers.com.

Manufactured in the United States of America

1 3 5 7 9 10 8 6 4 2

Library of Congress Cataloging-in-Publication Data
Names: Mosley, Michael, 1957 – author.
Title: Fast asleep : improve brain function, lose weight, boost your mood, reduce stress, and become a better sleeper / Dr. Michael Mosley.
Description: New York : Atria Books, [2020] | Includes bibliographical references and index.
Identifiers: LCCN 2020011971 (print) | LCCN 2020011972 (ebook) |
ISBN 9781982106928 (hardcover) | ISBN 9781982106942 (ebook)
Subjects: LCSH: Sleep—Popular works. | Sleep disorders—Popular works. |
LCGFT: Cookbooks
Classification: LCC RA786 .M68 2020 (print) | LCC RA786 (ebook) |
DDC 613.7/94—dc23
LC record available at https://lccn.loc.gov/2020011971
LC ebook record available at https://lccn.loc.gov/2020011972

ISBN 978-1-9821-0692-8
ISBN 978-1-9821-0693-5 (pbk)
ISBN 978-1-9821-0694-2 (ebook)

CONTENTS

INTRODUCTION

Sleep is something we all do; in fact, we spend around a third of our lives in this strange, unconscious state. And yet until recently we understood very little about what sleep is for, how much we need, and the role that dreams play in improving our mental health.

The good news is that over the last 20 years there has been a revolution in our understanding of sleep and just how important it is. Not so long ago it was fashionable to brag that you hardly slept at all, and the mark of a successful businessperson or politician was their ability to get by on very little shut-eye. Former British prime minister Margaret Thatcher was held up as a shining example of someone who could operate without much sleep (which turned out to be a carefully cultivated myth), while I remember being told by a grizzled medical consultant, when I complained about the impact that lack of sleep was having on my empathy and judgment, that "sleep is for wimps." Or, as another put it: "There's plenty of time to sleep when you are dead."

Our current attitudes toward sleep are very different. Thanks to recent research, we know that too little sleep can devastate your body, brain, and microbiome (gut bacteria), dramatically increasing your risk of developing a range of chronic conditions such as obesity, type 2 diabetes, and dementia.

And, when it comes to sleep, it's not just about quantity, but about quality, too. We have learned, through extensive sleep studies, that if you don't get enough of the right sort of sleep, you increase your risk of depression and memory problems. Which is all very worrying, particularly if, like a third of the adult population, you suffer from insomnia.

Fortunately, there are surprising and highly effective ways to improve your sleep quality, ensuring you fall asleep rapidly, get plenty of deep sleep, and wake up feeling refreshed. This, in turn, should boost your happiness, creativity, and even life expectancy.

The reason I particularly wanted to write this book is because I am obsessed by sleep and have been for many years, not just from a science perspective, but also on a deeply personal level. For the last 20 years, I have suffered quite badly from intermittent insomnia, to the point where I was in real despair. I wanted to find out what I was doing wrong and, of course, I wanted to find out what I could do to make it better.

I wasn't always a poor sleeper. When I was a teenager, I could sleep anytime, anywhere. I once slept in a photo booth (I had missed the last train home). Another time I slept in a telephone kiosk. I never worried about going to sleep or staying asleep, because that came naturally.

I didn't always get a good night's sleep, but that was my choice. Like most teenagers, I was keen to burn the candle at both ends. As a medical student, I often stayed up partying, then went straight into some feverish last-minute cramming. Which I now realize was wildly counterproductive. You need sleep to consolidate your memories, as I will explain in this book.

As my medical training progressed, sleep became ever more precious. I found I just couldn't function any longer on a few hours' sleep a night. I became intensely irritable and I'm sure that both my judgment and my empathy were impaired. But, even so, I could still go to sleep and stay deeply asleep for hours when I was given the chance. Despite the disruption to my sleep pattern caused by the irregular hours I had to work, I never had any problem drifting off.

Then, as I entered my late twenties, everything changed. By then I was married and I had started a new career in television. The hours were long and unpredictable, though nothing like as bad as in medicine. At this time my wife, Clare, was working as a junior doctor and regularly working 120 hours a week. It was not unusual for her to be on duty for three or four days with only a few hours of broken sleep a night, which blunted her thinking. She told me that after one particularly grueling week, she briefly fell asleep standing, during an operation. Fortunately, she woke up before anyone noticed.

Not only did work absorb almost every waking hour, it also began to intrude on our sleep. On the occasions that Clare was actually sleeping at home, she would regularly wake me up in the middle of the night to help her look for patients, which in her sleep-deprived state she was convinced were lost in the cupboards or waiting for her downstairs. Clare has parasomnia, a quite common set of sometimes bizarre nocturnal behaviors, which include a tendency toward sleepwalking and sleep talking.

By the early 1990s, we had started to have children, and that, of course, resulted in many nights of disrupted sleep. In fact, we went on to have four children, which meant that a full decade was dominated by babies.

By the time we entered our forties, Clare was a GP and working more regular hours. Our children were also sleeping through the night. But by then I had begun to show classic signs of insomnia. I had difficulty going to sleep and kept waking up at three in the morning with thoughts rushing through my head. I would lie there for what felt like hours, and going to bed, which was once a real pleasure, was something I began to approach with a sense of unease. Would this be a good night or a bad night? Would I get up feeling shattered or would this be one of those rare nights when I would sleep through until morning?

Naturally enough, I wanted to understand what was going

on and what I could do to get back to the days of blissful, effortless sleep. I made what was to be the first of many popular television programs examining the mystery of sleep. Making these programs introduced me to lots of sleep scientists and a whole new, fascinating world of sleep research.

To try to understand the impact of severe sleep deprivation, I decided to see how long I could stay awake with a man who holds the unofficial world record. He can go days on end with no sleep without appearing to suffer. What was the secret to his success? Why could he just keep going, while I couldn't?

Since then I have spent many nights in sleep labs with electrodes attached to my head and body. I've taken drugs to put me to sleep and drugs to keep me awake. I have interviewed hundreds of people, ranging from firefighters to doctors, astronauts to police officers, about their sleep. I have also begun to look at the impact of food on sleep and test out different ways to improve sleep quality.

The Structure of the Book

You may be someone who is desperate to get a good night's sleep. Or you may simply be interested in what happens to you when your eyes close and you drift off into the land of Nod.

The first part of this book is all about the science of sleep: the research that has led to our current knowledge and how this has given us rich insights into a previously undiscovered land. What are common sleep disorders and how do they arise? What really happens to your brain and body when they are chronically sleep deprived? Why are dreams so important and how can you make the most of them?

I will use my own sleep adventures to illuminate the journey and I will, of course, provide plenty of scientific studies to justify my more surprising claims.

All of this lays the groundwork for the second part of the book, which is primarily aimed at helping you sleep better. After all, I suspect that many of you are reading this book because you suffer from occasional insomnia, or you know someone who does.

I will take you through the best that modern science can offer with a sleep program that should, within a few weeks, set you on a better path.

One of my key goals is to help you improve your "sleep efficiency," which is a measure of how well you've slept. Your sleep efficiency represents the amount of time you spend in bed fast asleep, as opposed to trying to get to sleep or lying in bed wide awake, fretting. You should be aiming for a sleep efficiency of 85%. More on that later.

As for the Fast Asleep program, it has been put together based on well-tested scientific principles. Yet at its heart are two novel and surprising elements, both based on the latest scientific research.

The first thing that might surprise you is that the most effective way to cure insomnia is to reboot your brain by putting yourself through a short course of Sleep Restriction Therapy. It is called Sleep Restriction Therapy because, paradoxically, it demands that you cut back on your sleep. Yes, I am going to help you sleep better by asking you to cut the amount of time you spend in bed. This can be demanding and is not an approach that is going to be appropriate for everyone, so if you have any health issues or are concerned about your suitability, then do talk to your doctor before beginning.

One of the classic mistakes people who have problems sleeping make is to try to spend *more* time in bed—when, for most people, lying in bed not sleeping isn't restful, it is very stressful. It also sets up a really bad behavior pattern where your brain comes to associate being in bed with being awake, fretting.

Studies have shown that sleep restriction is more effective than anything else, including drugs, and that the results last, long-term.

The second novel thing about my program is the emphasis I place on food, particularly the sorts of foods that have been shown to improve the quality of sleep. Forget all those stories about turkey or cheese. It turns out that eating more legumes and fiber-rich foods, and fewer late-night sugary snacks, is one of the most effective ways to boost your levels of deep sleep and improve your mood.

That's, in part, because fiber-rich foods feed the trillions of "good" bacteria that live in your gut, which, in turn, produce chemicals that have been shown to reduce stress and anxiety. I will take you on a fascinating guided tour of the science.

On a more practical level, my wife, Clare, along with food writer Justine Pattison, has created a range of tasty recipes that are packed with the sorts of ingredients that those good bacteria will really love, and that you will love, too.

I do hope you enjoy this book and, most of all, I hope it puts you to sleep. Fast.

1.

HOW WE WOKE UP TO SLEEP

As I pointed out in my introduction, it is astonishing that despite the fact that we spend up to a third of our lives—about 25 years—asleep, until relatively recently we knew very little about what went on during that time. A hundred years ago, most people thought that the brain simply switched off, like a light bulb, when you went to sleep.

The American inventor Thomas Edison, who manufactured the first light bulbs, and whose invention did more than any other to disrupt our sleeping patterns, thought that sleep was a waste of time. He claimed to need less than five hours' sleep a night and said that having more was just greedy. As he put it: "Most people overeat 100%, and oversleep 100%, because they like it. That extra 100% makes them unhealthy and inefficient."

As we'll see, he couldn't have been more wrong. One reason we knew so little about sleep was that until the early 20th century we had no means to probe it. Scientists like being able to measure things, and sleep was all too intangible. It was like trying to make sense of the movements of the planets before we had the ability to properly study the heavens.

The man who made the first major breakthrough in the science of sleep was a peculiar German psychiatrist named Hans Berger.

Reading the Sleeping Mind

Hans Berger's contribution was the invention of electroencephalography (often abbreviated to EEG), the ability to record human brain waves by attaching electrodes to a volunteer's skull.

He built the first working electroencephalograph in 1924, but for a long time his work was ignored. He was widely regarded as a crank. And that's not surprising because Berger believed passionately in telepathy. In fact, the main reason he had created his EEG machine was to prove that humans can communicate with each other through psychic powers.

His obsession with telepathy began when he was a young

cavalry officer. One day, while he was taking part in a military exercise, his horse suddenly reared and threw him into the path of a horse-drawn cannon. He wasn't seriously injured, but he was very shaken. He later discovered that his sister, who was at home at the time, had had a sudden premonition that he was in deadly danger and forced their father to send him a telegram to see if he was alright.

Berger was convinced that during the accident, he had sent out a powerful psychic distress message, which his sister had somehow picked up. He was so convinced of this that he decided to retrain as a doctor, and then as a psychiatrist, just to prove that telepathy exists.

I personally don't believe in telepathy, but Berger was absolutely right when he claimed that the human brain produces electrical signals that can be "read" by multiple electrodes placed on the scalp. Although modern versions of the EEG are far more sophisticated than Berger's invention, in essence they do the same thing.

Dream Sleep

Berger had shown, in 1924, that his EEG machine could be used to study human brain waves, but it would be another 27 years before sleep researchers got around to using it in any meaningful way.

In December 1951, an impoverished student at the University of Chicago named Eugene Aserinsky decided to take his eight-year-old son, Armond, to his lab to take part in a novel sleep experiment. He scrubbed Armond's scalp, taped on the EEG electrodes, and left him to fall asleep. Aserinsky then went next door to record what happened.

For Aserinsky this was a do-or-die moment. He was 30 years old and living with his newly pregnant wife in a converted army barracks. They were so poor he could barely afford the HP payments on his typewriter, let alone heat their home. He needed a breakthrough with his research, and soon.

Since no one else had yet bothered to use an EEG to study someone sleeping through an entire night, Aserinsky had decided he might as well begin with his young son.

Nothing very exciting happened for the first hour, but then he noticed that his machine had begun to record a sudden change in brain activity. On the machine, it looked as if his son Armond had woken up. But when Aserinsky went next door to check, Armond was clearly still deeply asleep. Nothing was moving except for his eyes, which were darting around under his eyelids.

Aserinsky woke the boy, who reported that he had been having an intense dream. This was amazing, groundbreaking stuff. The next day Eugene repeated the experiment, with the same results. A few hours after Armond fell asleep, the EEG recorded a sudden change in his brain activity, which coincided with rapid eye movements. Studies done with other, adult, volunteers showed the same thing.

Eugene Aserinsky had done something that would transform our understanding of sleep. He had sent the first exploratory probes into Planet Sleep and discovered that, far from being a dull, barren world where not much happens, it is a place where the brain undergoes some remarkable changes. Sleep research was about to become sexy.

However, despite having made this amazing breakthrough, Aserinsky soon lost interest in sleep. After publishing his findings in 1954, he went off to study electrical brain activity in salmon and later died in a car crash, very possibly because he fell asleep at the wheel.

So what does happen when we sleep?

I've spent many nights both being observed in sleep labs and, more interestingly, observing other people while they sleep. If you have never seen someone else go to sleep, or had yourself filmed while falling asleep, then I would recommend you give it a go. It is very entertaining.

As I mentioned earlier, we used to think of going to sleep as being like switching off a light bulb. You were either awake or asleep. We now know it is much more complicated than that.

Sleep involves three distinct states: light sleep, deep sleep, and REM (rapid eye movement) sleep. We sleep in roughly 90-minute cycles throughout the night, flipping between one state and another.

As you can see from the diagram opposite, during the first part of the night you get most of your deep sleep, while the second half of the night is dominated by REM sleep. Most people wake two or three times a night. If you are lucky (like my wife, Clare), you won't even be aware of it. If you are unlucky, you will wake up and stay awake.

When you go to bed and close your eyes, you should soon start to drift off into light sleep (Stage 1). At this point you are drowsy, but easy to rouse. If the dog next door starts to bark or your partner starts to snore loudly, you might wake up.

Hypnogram

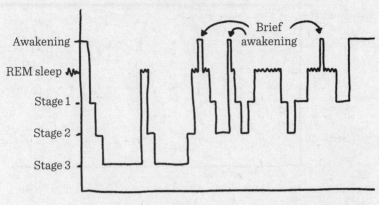

After Stage 1 (which normally lasts about ten minutes), you begin a deeper dive into sleep.

When it comes to sleep, I like to see myself as a seal, plunging joyfully into the depths of the night. Some years ago, I made a film about free divers, people who swim to great depths without the aid of oxygen tanks, and it looked just beautiful, as they headed away from the bright light of the surface toward the dark of the ocean floor. That said, for some people, the act of falling asleep is frustrating rather than joyful.

The next stage, Stage 2, also counts as "light sleep." As you enter it, your core body temperature (typically measured with a rectal thermometer), which began to fall even before you got into bed, drops even farther. Your heart rate slows down (I've recorded mine and it drops from its usual 60 beats a minute to about 55) and your breathing also slows and becomes steadier.

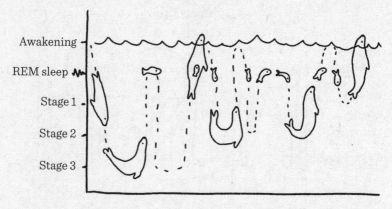

When you are entering Stage 2, you may well do what is known as a hypnagogic jerk or "sleep start." This is an involuntary twitch of the muscles as you go deeper into sleep. Most of us do it. Although it is normally no more than a little twitch, some people really lash out, which is no fun if you are sharing a bed with them. It is often a sign of stress, and if you follow the regimen I'm going to outline in this book, you should not only sleep better but also be less likely to do this annoying little dance at night.

If all is going well, within an hour of beginning to nod off you will have entered Stage 3—deep sleep. This is also known as slow-wave sleep because it is when an EEG would start recording slow, deep waves. Millions of neurons in your brain fire at once, then pause, before firing again. This activity creates great, crashing waves that travel through the brain, which are really hypnotic to watch on a screen.

In deep sleep you are at your most relaxed and difficult to rouse. But, while your brain rests, your body is hard at work

14

because deep sleep is when a lot of vital repair work gets done. Your pituitary gland, for example, will start to secrete more growth hormone, which is vital for cell growth and repair. Deep sleep also boosts your immune system.

Without sufficient deep sleep, your body makes fewer cytokines, a type of protein that regulates your immune system. Cytokines are vital for helping you fight infections, which is why lack of sleep makes you more vulnerable to catching colds and also reduces the efficacy of vaccines against diseases such as the flu.

Despite the fact that deep sleep is supposed to be a time of deep relaxation, it is also the time when some people start to do strange things, like sleepwalking, sleep talking, and even sleep eating. I will talk about these phenomena in more detail in Chapter 2.

Deep Sleep and Brain Cleaning

When I was young, I loved reading stories from Greek mythology and one of my heroes was the super-powerful Heracles (known as Hercules by the Romans). Heracles, who was the son of Zeus, was told he could be made immortal if he successfully carried out 12 apparently impossible tasks ("the labors of Heracles").

The least glamorous of these challenges was to clean out the Augean stables, in a single night. The Augean stables were notorious because they housed more than 3,000 cattle and hadn't been cleaned for years. You can just imagine the stink. Yet Heracles succeeded in scouring the stables of decades of accumulated dung, in a single night, by diverting two rivers through them.

The reason I bring up this story is that, overnight, something similar takes place inside your head. A network of channels in your brain, known as the glymphatic system, opens up and pumps cerebrospinal fluid through it while you are in deep sleep. Like the rivers in the Augean stables, this fluid flows through your brain tissue and washes away the toxic waste that has built up there during the day.

That's the good news. The bad news is that, as we get older, we tend to get less deep sleep, which means that our brains are not as good at washing away the toxins. Young people typically get a couple of hours of deep sleep a night. When you get to my age (63), you are lucky if you are getting 30 minutes.

This matters because it is the accumulation of toxic proteins in the brain, such as beta amyloid and tau, that appears to drive Alzheimer's disease, and in humans there is a very clear link between poor sleep and the development of dementia.

To maximize your chance of getting deep sleep, it is a good idea to go to bed before midnight, since your brain gets the most deep sleep during the first half of the night. Eating the right foods has also been shown to boost deep sleep, which I will discuss further in Chapter 5.

Deep Sleep and Memory

As well as giving your brain a good spring clean, deep sleep is when your brain sorts out your memories and shifts the useful ones into deep storage.

During an average waking day, an awful lot happens to you. You listen to the news, read a book, go to work, talk with friends, go on social media, listen to music. In other words, you load your brain with a myriad of potential memories. Some are useful, but others can be happily discarded. It is while you are asleep (particularly in deep sleep) that your brain decides which memories it wants to keep and which to discard.

It's a bit like sorting out photos and videos on your phone. Storing images requires a lot of memory, so when your phone starts to get full, you have to edit them. Removing dud videos and photos leaves space for new ones.

Even compared to a modern computer, your brain can store an extraordinary amount of data; a recent estimate puts its storage capacity at about 1,000 terabytes, which is a billion megabytes. A computer with that capacity could store around 2 billion books or 500,000 films.

Yet while you have an awesome capacity to remember things, you don't want to store more junk up top than necessary. So, during the night, the memories that are considered important are shifted from the hippocampus (the short-term storage area of the brain) to the safety of the prefrontal cortex (the long-term storage area of the brain—think of it as your hard drive). The memories left behind in short-term storage are gradually deleted.

That's why, if you are a student, getting a good night's sleep before an exam is so important. Staying up late and cramming is self-defeating because all those last-minute facts being madly forced in will soon be gone. You might think: "I'll cut back on sleep during the week and then make up for it on the weekend." Unfortunately, it doesn't work like that, because memories need to be consolidated within 24 hours of being formed.

A dramatic fall in the amount of deep sleep we typically get as we age may also explain why our ability to remember things gets worse as we get older.

In a recent study, researchers from the University of California, Berkeley,[1] asked 18 healthy young adults (mostly in their twenties) and 15 healthy older adults (who were mostly in their seventies) to come into the sleep lab to take part in a memory test. Before going to bed, they were asked to memorize pairs of words, and were tested to see how well they did.

They were then attached to an EEG machine, which measured their brain wave activity while they slept. The next morning, they were tested again to see how many of the word pairs they could recall.

The researchers found that the older participants got 75% less deep sleep than the younger participants, and their ability to remember word pairs was 55% worse.

Brain scans also showed that, overnight, the youngsters were much more efficient when it came to shifting memories from the short-term storage of the hippocampus to the long-term storage of the prefrontal cortex.

One encouraging finding was that applying "transcranial direct current stimulation"— a small electric buzz to the surface of the brain—enhanced deep sleep in the older participants and improved their ability to do well in the memory test. Even so, as you'll discover in Chapter 6, there are easier ways to enhance deep sleep than giving your brain electric shocks.

REM Sleep and Emotions

As we've seen, deep sleep is vital for cleaning out our brains and sorting our memories. REM sleep, which occurs later in the night, is also important for tidying and organizing our memories, but it has the additional role of helping resolve our emotional issues.

Although we dream at other points in the night, REM sleep is when we have our most vivid dreams, and it is these dreams that help us process and deal with bad memories and experiences. All of which helps explain another very odd finding: during REM sleep most of our muscles are paralyzed. This is probably so that while in the grips of an intense, dramatic dream, we don't thrash around and hurt ourselves. We do go on breathing, taking short, shallow breaths, but apart from that, the only part of us that is obviously moving is our eyes.

If you look at someone in REM sleep, you will see that, underneath their eyelids, their eyes are flicking madly to and fro. No one knows why this happens, but one theory is that it reflects the sort of eye movements you might make while watching a film. Dreams have been called the cinema of the mind, so perhaps the eye movements are simply a sign that you are following the action.

So how does REM sleep help us process our emotions? Well, it is all to do with the amygdalae, the two almond-shaped groups of cells located deep in the brain that play a key part in regulating emotions. Let's first look at how they work while we are awake.

I am mildly claustrophobic, and when I am in a confined

space, I start to feel a sense of rising panic. That is because my amygdalae have triggered the release of "fight or flight" hormones, such as adrenaline, and this, in turn, makes my heart rate, blood pressure, and breathing shoot up. I feel nervous, sweaty, and sometimes nauseous. There is a part of me that knows nothing bad is going to happen, but most of me just wants to escape from the situation.

Since the release of "fight or flight" hormones plays such a big part in generating fear responses, I was fascinated to discover that REM sleep is the one time of day or night when links to these stress-inducing chemicals in the brain are switched off. This means that, while the dreams we have during REM sleep can be scary and disturbing, they are not nearly as bad as they would be if you were having them while you were awake.

Looked at this way, dreaming during REM sleep is a form of psychotherapy, where you revisit unpleasant memories and events but remain calm. This allows you to process your emotions and defuse them.

• •

The Spider Dream

While writing this book, I asked lots of people about their sleep and their dreams. The following story, which someone told me, is a great example of a therapeutic dream: "When I was young I had a fear of spiders; it wasn't terrible but I had to leave the room if I saw one. Then one night I had a dream in which I was sitting on a chair in a dark room. From the chair I could see a door. There was a light under the door and I noticed that small spiders were crawling through the gap under the door. Slowly the gap under the door got bigger and bigger and as it did larger and larger spiders started coming through. For some reason I wasn't scared, I was just curious to see how big they would get. Then I woke up. The oddest thing is that after I had that dream, I wasn't scared of spiders anymore. In fact the next time I saw one I was able to pick it up without shrieking."

Sleep Your Way to the Top

One other great thing about REM sleep is that it makes us more creative. It seems that the age-old advice about "sleeping on a problem" is spot on: research has shown that a good night's sleep, particularly one that is rich in REM, increases our ability to come up with novel solutions to problems.

When I am struggling with a problem, I often write it down in a notebook, put it aside, and then go back to it the next morning. I find it really helps. Decisions made late at night, or after a bad night's sleep, are often the ones we regret.

There are lots of lovely stories about people who have had their eureka moments while they slept.

- The writer Mary Godwin (later Mary Shelley) came up with the idea of Frankenstein after having a dream about a scientist who created life and was horrified by what he had done.
- Paul McCartney says the tune for "Yesterday" came to him while he was asleep.

- Even more impressively, Keith Richards says he not only dreamed the opening lines to what would become one of the Rolling Stones' greatest hits, but played the song in his sleep. The story goes that he often kept a guitar and tape recorder by his bed, and one morning in May 1965, while on tour in Florida, he woke to find that the recorder had been running during the night. When he played it back, he heard himself playing the opening verse to "Satisfaction."

- Since REM sleep is all about excitable neuron activity in the brain, it is appropriate that the scientist Otto Loewi, who first showed us how nerves communicate, made his remarkable breakthrough thanks to a dream. In the spring of 1920, Dr. Loewi was a frustrated man. He was convinced that nerve messages were transmitted using chemical signals, but he had spent 17 years trying to prove this and failed. Then, during the night of Easter Sunday that year, he had a dream. He woke and jotted down a few notes on a slip of paper, before falling asleep again. When he woke up the next morning, he remembered he had written down something important, but when he picked up the paper he couldn't read his own handwriting. Nor could he remember anything about the dream. Fortunately, the next night he had the same dream. This time he woke up properly and wrote the whole thing down. In the dream he was performing an experiment on frogs that would allow him to test his theory. As he later wrote: "I got up immediately, went to the laboratory, and performed a simple experiment on a frog heart according to the nocturnal design." The experiment worked and later won him the Nobel Prize in Medicine.

Things That Go Wrong If You Don't Get a Decent Night's Sleep

Why Lack of Sleep Makes You Fat

A bad night's sleep not only affects your brain, but also messes with your body, including its ability to control your blood sugar levels. In the long run, this can lead to obesity and diabetes.

A few years ago, to see what impact even a couple of nights of reduced sleep can have, I took part in an experiment with Dr. Eleanor Scott, who works at the University of Leeds. We recruited a group of healthy volunteers and fitted them with activity monitors and continuous glucose monitors—devices that are strapped to the arm to measure blood sugar levels. In this way, we were able to constantly monitor what was happening to their blood sugar levels without having to prick their fingers repeatedly.

First, we asked our volunteers to sleep normally for two nights (so we had a baseline to work from), then go to bed three hours later than normal for two nights.

I felt I couldn't ask our volunteers to do this experiment unless I was ready to do it myself. I was also curious to see what effects it would have on my blood sugar levels. Back in 2012, I had discovered that I had type 2 diabetes, which I managed to get rid of by putting myself on the 5:2 diet and losing 20lb (9kg). Would a couple of nights' bad sleep set me back, even now?

After two nights of severe sleep deprivation, I went back up to Leeds, where I met up with Dr. Scott and the other volunteers. Everyone complained about having the munchies.

As one of them put it: "I wanted lots of biscuits and I didn't just have one. I'd go for 10 custard creams."

"Is that unusual?" I asked him.

"Well, it's certainly unusual for breakfast!"

All of us, whether we had feasted on cookies or managed to stick to our normal diet, saw marked increases in our blood sugar levels when we were badly sleep deprived, to the point where some of us (myself included), who had normal levels at the start of the experiment, now had those you would expect to see in a type 2 diabetic. Our blood sugars returned to normal after a good night's sleep.

As Dr. Scott pointed out, there is now a lot of evidence that people who sleep badly most nights are far more likely to become overweight or obese and develop type 2 diabetes than those who sleep really well.

So why does this happen? "We know that when you are sleep deprived," Dr. Scott said, "this alters your appetite hormones, making you more likely to feel hungry and less likely to feel full. We also know that when people are sleep deprived, they often crave sweet foods, which could explain the custard cream cravings. Also, if you're awake when you're not meant to be, you produce more of the stress hormone cortisol, and that can influence your glucose level as well the next day."

Ours was quite a small experiment, but a recent meta-analysis, carried out by researchers at King's College London,[2] found that sleep-deprived people consume, on average, an extra 385 calories per day, which is equivalent to a large slice of cake.

It's not just that your blood sugar levels soar and your hunger hormones go into overdrive when you're tired; the areas of your brain associated with reward also become more active. In other words, you become much more motivated than normal to seek out unhealthy foods such as chips and chocolate.

Another study[3] showed that children are affected in a similar way. Researchers took a group of children aged between three and four years, all regular afternoon nappers, and not only deprived them of their afternoon nap but also kept them up for about two hours past their normal bedtime.

The following day, the children ate 21% more calories than usual, including 25% more sugary snacks. They were then allowed to sleep as much as they wanted. The next day, they still consumed 14% more calories than they had before being deprived of sleep.

The Vicious Circle

Lack of sleep makes you fatter, but piling on extra fat (particularly around the gut and neck) also means you sleep worse. It is a vicious circle. When I was an overweight diabetic, I slept badly, at least in part because I snored so much. Being overweight also greatly increases your risk of having sleep apnea, a disorder that causes you to stop breathing hundreds of times a night. This will make you really tired and hungry and it is terrible for the brain.

A Swedish study[4] showed just how disruptive being overweight is when it comes to sleep. For the study, the researchers recruited 400 women with an average age of 50 from the town of Uppsala. Half the women were overweight and had "central obesity," meaning their waists measured more than 35 inches (88cm). The other half were slimmer, with a body mass index (BMI) in the normal range. After being weighed and having their waists measured, the women were led away to the sleeping area, where the team hooked them up to sleep recorders.

The women were then allowed to sleep for as long as they wanted. The differences in sleep quality and quantity were striking. The women with less belly fat slept, on average, 25 minutes more per night. They also got 20% more brain-restoring deep sleep and 22% more emotionally calming REM sleep.

Metabolic Syndrome

Lack of sleep also contributes to metabolic syndrome, which is the medical term for a cluster of conditions that includes too much body fat around the waist and raised blood pressure, blood sugar, and cholesterol, which, in turn, lead to an increased risk of type 2 diabetes, stroke, and heart disease. Metabolic syndrome, also known as Syndrome X, affects about one in four adults in the UK and has a major impact on future health, not only because it encourages further buildup of fat, particularly around your gut (visceral fat), but because it leads to increased insulin resistance. In other words, your body has to pump out ever-increasing amounts of insulin to bring your blood sugars back to normal.

Low Mood

Anyone who has been sleep deprived knows that it leads to anger and irritability, while at the same time sucking the joy out of life. If you are feeling anxious and depressed this will, in turn, affect how well you sleep. Being agitated keeps your body and brain aroused, just when you want them to wind down. More on how to combat this in Chapter 4.

How lack of sleep impacts your health

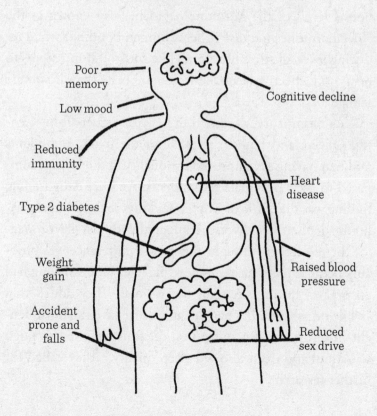

Poor memory

Low mood

Reduced immunity

Type 2 diabetes

Weight gain

Accident prone and falls

Cognitive decline

Heart disease

Raised blood pressure

Reduced sex drive

Sex Drive

As well as making you feel simply "too tired for sex," being sleep deprived suppresses the production of the two main sex hormones, estrogen and testosterone. This, in turn, has a devastating effect on sexual desire.

The good news is that getting more sleep should improve your sex drive. A two-week study[5] of 171 American women found that an extra hour of sleep increased the chance that they would want to have sex the following night by 14%.

But does it work the other way around? Will having more sex improve your sleep? In theory it should, as regular sex boosts levels of the hormone oxytocin (also known as the "love hormone" because it helps human bonding) while reducing levels of stress hormones like cortisol. But it seems to be a more effective sleep aid if you are male than if you are female.

In a recent survey[6] carried out by researchers from Central Queensland University, 68% of the male respondents said that having sex improved the quality of their sleep, compared to 59% of female respondents. Another enlightening finding was that 11% of women said that sex with their partner made their sleep worse, compared with just 4% of men.

The researchers thought this gender difference was probably because the men were more likely to report having had an orgasm during sex than the women. They added that "while orgasms with a partner appear to have the most benefit in terms of sleep outcomes, orgasms achieved through self-stimulation can also aid sleep quality." They called for further research . . .

Keeping Track of Your Sleep

It's all very well being told that getting enough sleep is vital for your brain, your waist, and your sex life, but how do you track how much you are getting? Most people don't have access to a sleep lab.

Without an EEG machine you won't get a very accurate picture of your sleep, but some of the more modern sleep trackers, which also measure your heart rate, do a reasonable job. I spent some time looking into trackers and, in the end, I went for the Fitbit Alta HR Activity Tracker with Heart Rate Monitor.

The Fitbit Alta HR has been the subject of proper clinical trials, including a recent study[7] where researchers from Monash University in Australia compared the accuracy of the tracker with the data collected from 49 people linked to machines in a sleep lab. The researchers concluded that the watch was pretty good at telling when people were in deep sleep and REM sleep, but tended to overestimate how much sleep people got because it was not great at detecting when people were awake but not moving.

That said, I do think it is worth using a sleep tracker. And I recommend that you keep a sleep diary as well.

A Sleep Diary

I wore the tracker for several months, while keeping a detailed sleep diary, and the tracker did seem to reflect how much I had slept and how well I had slept. I have included a copy of a daily page from a sleep diary in Chapter 6, and you can download more copies at fast-asleep.com. If you are curious about your sleep, and certainly if you suffer from any form of insomnia, then keeping a sleep diary is an absolute must. I will go into the whys and wherefores later.

Sleep Efficiency

One of the other things that a sleep tracker will allow you to do is calculate your sleep efficiency, which is a core part of the Fast Asleep program. What is sleep efficiency? As I wrote earlier, it is a measure of the amount of time you spend in bed actually asleep.

Let me illustrate with my own example.

I try to be in bed by 11 p.m. and get up at 7 a.m. most days of the week, including weekends. Routine is hugely important if you want to sleep well, and I value my sleep.

My sleep tracker told me that over the course of a month I was in bed for an average of 7 hours and 50 minutes a night, but I was only asleep for 6 hours and 40 minutes. The rest of the time I was trying to sleep, roaming the house, or reading.

If you translate that into minutes, you can calculate my sleep efficiency:

$$(6 \times 60) + 40/(7 \times 60) + 50 = 400/470 = 85\%$$

85% is actually pretty good. In fact, anything between 85% and 90% is excellent, and very few people have a sleep efficiency that is over 90%. An insomniac will probably spend about 70% of the night asleep.

And what about the different stages of sleep? According to my tracker, of the 6 hours and 40 minutes that I was asleep, I was in deep sleep for 17% and in REM for 18% of the time. The rest of the time I was in light sleep.

Which isn't bad. According to a recent Fitbit study,[8] the average user lies awake for 55 minutes every night and sleeps for about 6 hours and 33 minutes. They are in deep sleep for 15% and in REM for about 20%.

Again, according to Fitbit, the older you are the less sleep you tend to get, down from 6 hours and 58 minutes for young adults to 6 hours and 33 minutes for people aged 52 and older. The big drop was in the amount of deep sleep people got as they aged, down from 71 minutes a night to just 50 minutes for the over 50s.

In the Fitbit study, the women got slightly more sleep than the men, by about 25 minutes a night, and this is reflected in other studies of American adults.

It seems that women put a higher priority on sleep, perhaps because they are better at recognizing the health benefits. There is also evidence that women are more vulnerable to the impact of sleep deprivation.

Summary

- There are three distinct states of sleep: light sleep, deep sleep, and REM sleep.
- During the night we sleep in roughly 90-minute cycles,

with most deep sleep happening in the early part of the night, and most REM sleep happening later on.

- About half the night is spent in light sleep.
- Deep sleep is when your brain gets cleaned and your memories get sorted.
- REM sleep is when you have vivid dreams that help you process your emotions.
- Not getting enough sleep increases your risk of obesity, type 2 diabetes, dementia, high blood pressure, and low mood. It also reduces your sex drive.
- The really important thing is not how long you spend in bed but how long you spend in bed ASLEEP. Being able to estimate your sleep efficiency is key to doing the Fast Asleep program.

2.

WHAT PUTS US TO SLEEP AND WHAT KEEPS US AWAKE

The urge to sleep begins early in the day, from the moment that you first wake up and get out of bed. Shortly before you wake, your body releases a surge of hormones, including the stress hormone cortisol, which prepares you for the day. But waking also triggers the release of a chemical in the brain called adenosine.

Adenosine binds to receptors in your brain, slowing down brain activity. This suppression of brain cell activity is what causes the feeling of drowsiness.

The longer you are awake, the higher your adenosine levels rise. The higher the adenosine levels, the sleepier you get. Then, when you finally go to bed and sleep, the adenosine is broken up and disposed of.

If you are feeling sleepy and want to stay awake, you can, temporarily, block the effects of adenosine by consuming the world's most popular psychoactive drug, caffeine. Caffeine binds to sleep-inducing receptors in your brain that would otherwise be occupied by adenosine—that's why it wakes you up.

As you may have noticed, caffeine has very different effects on different people. There are some people, like my wife, who are really sensitive to caffeine. She only needs one cup to wake her up. Caffeine also hangs around in her system for much longer than it does in mine.

The reason you have to keep topping up your caffeine is that it is constantly being broken down by your liver. The average half-life of caffeine is around five hours, which means that if you have a cup of coffee at 6 p.m., half is still running around your system at 11 p.m. and a quarter is still there at 4 a.m.

If you are sensitive to caffeine, you will still be feeling the effects of an afternoon cup of coffee in the middle of the night, lying awake and wondering why you can't sleep.

But quoting average figures is misleading because some people break down caffeine much faster than others. The "average" half-life of caffeine may be 5 hours, but the range is actually 1.5 hours to 9 hours. If you are someone whose body breaks down caffeine especially slowly, you will still be feeling the effects of your morning cup of tea or coffee in the middle of the night. But, if you are naturally able to break down caffeine rapidly, then you can drink coffee in the evening without it disturbing your sleep.

Various things affect the half-life of caffeine in your body, including your gender, age, weight, and any medication you are on. Taking the oral contraceptive pill will dramatically slow your liver's ability to break it down.

The biggest factor, however, is your personal genetics. If you are interested, you can get yourself genetically tested by companies like 23andme.com. You order a test online, and a few days later they send you a plastic tube. You spit in the tube and send it back. A few weeks later, you log in and see what they have found.

I did this a few years ago and I found the results fascinating. According to 23andme, their test suggests that I am unlikely to have developed a bald spot (true) or get dandruff (true) but I can probably smell the asparagus odor in my urine (also true). As for caffeine, it turns out that I am more sensitive than most and I would be well advised to drink less of the stuff. These days, I have two to three cups of coffee during the morning and rarely drink any after midday.

Your Internal Clocks

As well as the buildup of adenosine, the other major driver of sleep is your circadian clock. Deep in your brain there is a small group of cells called the suprachiasmatic nucleus (SCN). If I were to drill a hole between your eyebrows and carry on drilling into your brain until I hit the hypothalamus, and if I were then to stick an electrode down that hole, you would actually be able to hear the clock tick.

Oddly enough, your circadian clock doesn't follow a day that is exactly 24 hours long. Some people's clock runs fast, others' runs slow. If you have a fast clock then that means you are a lark—you like to get up early. If you have a slow clock then you are an owl, someone who likes to stay up late. The reason you don't go completely out of whack is that your internal clock is reset every day by light.

Rays of light from the sun hit receptors in the back of your eyes that have nothing to do with vision but instead are linked to the SCN. The SCN then sends out signals to other parts of your body, including your guts, letting them know that a new day has begun and it is time to get moving. It's a bit like shouting at the kids, "Wake up, breakfast will be on the table in 20 minutes."

Similarly, just as we like to have a warm house when we get up, the SCN raises your core body temperature before you wake, so you are ready to get going.

It will also, in the early hours of the morning, switch off the production of melatonin—a hormone connected to your brain clock that is released when it gets dark to help tip you into sleep (more on that later)—and switch on the release of the stress hormone cortisol.

What makes things a bit more complicated is the fact that each of our organs has its own clock, which is linked to the main clock but not necessarily driven by it. The clock in your liver, for example, is not reset by light but by when you eat. This is important, because when your biological clocks get out of sync with the outside world, and with each other, you are in trouble. Not only will you struggle to sleep, but you will get hungry, have problems controlling your blood sugar levels, feel run-down and tired, and find it hard to concentrate. It's called "social jet lag," because just like the jet lag you experience after traveling across multiple time zones, it leaves you feeling terrible.

Fortunately, you can retune your clocks and get your body back into sync with some relatively straightforward tweaks. These involve making sure you eat the right food, at the right time, and that you get exposure to strong enough light, again at the right time. That is what this book will help you do. Fix the clocks. Fast.

Larks and Owls

Some people leap out of bed in the morning, raring to go. Others need an alarm clock—preferably two—to get them off to work on time.

My wife will quite happily stay up working until the small hours, while I prefer to head for bed soon after 10 p.m. I'm a lark; Clare is an owl. I'm itching to leave a party by 11 p.m.; she is just warming up. We have different chronotypes, and there is good evidence that this is not just personal preference but has roots in our genes.

For a bit of insight into what mine say about my sleeping patterns, I dropped in on Dr. Simon Archer, who is Professor of Molecular Biology of Sleep at the University of Surrey. Among other things, Simon has conducted studies looking at links between an individual's genetic markers and how vulnerable they are to sleep loss.

I had sent him the raw data from my 23andme gene test and he had taken a close look. What did his team make of it?

"Well," he said, "the first thing that stood out is you have three genetic markers that would predict you are a morning-type person." He then added, "We also found a marker for increased risk of insomnia and a marker which is associated with poor sleep efficiency in people who are exposed to high levels of work-related stress. From what I've seen, I would predict that you often have disrupted, fragmented sleep during the night and that you need more sleep than most."

All of which is true. I get really grumpy and sleep badly when I am stressed.

So my gene test confirmed what I had long suspected: that I am one of those people whose circadian clock is a little fast, meaning that I like to get up early and go to bed early.

Can you tell if you are a lark or an owl without doing a genetic test? Try answering the following seven questions with a straightforward "yes" or "no."

1. Do you wake up bright and cheerful, without an alarm clock, by 7 a.m.?
2. If you go to bed at 10 p.m., do you swiftly fall asleep?
3. Do you always try to head to bed before midnight, even if you are on vacation?
4. Or do you have trouble falling asleep before midnight?
5. Do you need an alarm clock to drag you out of bed in the morning?
6. Do you find you don't need to eat until later in the day and are happy with just a coffee?
7. Given the chance, would you sleep in till 11 a.m. or later?

If you answered "yes" to the first three questions and "no" to the last four, you have strong larkish tendencies. If it was the other way around, you are more owlish.

Why Larks Turn into Owls and Back Again

When it comes to your chronotype, it is not just about your genes. Both your age and gender also play a role.

A big study of 25,000 people of all ages who were asked to fill in a ChronoType Questionnaire[9] found that while most children are larks (early chronotypes), they become progressively more owlish through the teenage years, hitting maximum "owlishness" at around the age of 20. They try to sneak their phones up to their bedrooms, stay up late chatting to their friends on social media, and are hard to rouse in the morning. They grunt at breakfast, refuse your healthy offerings, and instead buy junk food and energy drinks on the way to school to wake themselves up.

Yes, this behavior is really annoying but, to some extent, it's not their fault. When puberty comes along, it shifts the internal clock to a later setting, by an average of one to two hours. So an angelic child who was once quite happy to go to bed by 9:30 p.m. and get up at 7 a.m. (getting the necessary 10 hours in bed) is now a stroppy teenager who resents being sent to bed at all and resents even more getting up at 7 a.m., after having managed less than seven hours' sleep.

And there are clear gender differences. Girls, who tend to hit puberty earlier than boys, also start turning owlish earlier, reaching "maximum owl" at the age of 19, before slowly becoming more larkish. Boys, on the other hand, have a body clock that tends to get them to bed later and later, until they hit the age of 21 and have to fit into the adult world. Nonetheless, they tend to remain more owlish than women until they hit their fifties, when gender differences disappear.

This is a real pain for both parents and offspring, because it causes a lot of conflict, but that may actually be what Nature intended. When kids are young, it is vital that their parents nurture and look after them. But as they grow older, they need to start to assert their own identity, to prepare for life outside the parental home, where they will have to fend for themselves, just as our ancestors did when they left the safety of the tribe. Staying up late, with other teenagers, while the parents are all asleep may be Nature's way of bonding the next generation together.

The trouble is, the modern world doesn't allow a lie-in. If you are a teenager, you may have an internal clock telling you to stay up late and get up late, but you also have parents who are shouting at you that it is time to get dressed and go to school. The result is that most teenagers burn the candle at both ends and fewer than 25% get the recommended nine hours during schooltime.

The obvious solution would be to start the school day later, and lots of schools have tried this. In 2016, the city of Seattle announced that most of its middle schools and all its high schools would start an hour later—putting arrival time back from 7:50 a.m. to 8:45 a.m. Parents were not keen and all the extracurricular schedules had to be changed, so there was a lot of grumbling from the staff. But was it worth it?

Well, researchers asked 170 students from two of the schools who were about to make the change to wear activity monitors to track their sleep. They found that putting schooltime back by almost an hour did indeed have a big effect on the amount these students slept. It went up from an average of 6 hours and 50 minutes before the change, to 7 hours and 24 minutes afterward, an added 34 minutes of sleep each night. Academic performance and school attendance also improved impressively, in line with getting more sleep.

Another study,[10] this time in Fairfax, Virginia, showed that delaying the start of the school day can cut car accidents. They found that students aged between 16 and 18 who drove themselves to school were nearly 9% less likely to have a crash if they went to a school where the day started later.

Turning an Owl into a Lark

Being a night owl leads to fights with your parents when you are a teenager and can be extremely inconvenient as you get older.

I recently met a night owl named Marie, who works full-time and has two young children aged three and five. She has been a night owl since her teens and she told me that she would, given half a chance, stay up until 2 or 3 in the morning and wake up at about 11 a.m.

Because she has young children she can't sleep in, and anyway, she has a job that involves a 9 a.m. start. So she goes to bed around 11 p.m. most nights, but then lies there awake until 2 a.m. or later.

She is rudely awoken by her alarm clock at 7 a.m., or by her children clamoring to be let into the bedroom. Her husband, who is also something of an owl, is able to function perfectly fine on five or six hours' sleep a night. But Marie can't. She told me that she wakes up most mornings feeling absolutely shattered.

She has tried all the obvious things, such as exercise and not lying in on weekends, but nothing works, except for sleeping pills, and that is not a road she wants to go down in the long term.

The good news for people like Marie is that it is possible, in just three weeks, to turn yourself from an owl into a lark, and without drugs.

You do it by resetting your internal clocks. And you do that by controlling exposure to light and the timing of your meals.

To show it can be done, researchers from Monash University in Australia recruited 22 night owls, men and women who normally go to bed around 2:30 a.m. and wake up at 10 a.m.[11]

For three weeks they were asked to follow nine simple rules. These were:

1. To wake up at least two hours earlier than normal, which for this group meant getting up by 8 a.m.
2. To get outside and expose themselves to plenty of outdoor light in the mornings.
3. To have breakfast as soon as convenient.
4. To only exercise in the morning.
5. To have lunch at the same time every day.
6. To avoid all caffeine after 4 p.m.
7. To avoid having a nap after 4 p.m.
8. To avoid bright lights during the evening and to head to bed a couple of hours earlier than normal—i.e., by about midnight.

9. To stick to this regimen every day of the week, including weekends.

After three weeks, the owls had successfully shifted their body clocks forward by an impressive two hours. Not only were they going to sleep earlier, but tests showed that their levels of melatonin, the sleep-inducing hormone, were peaking two hours earlier.

Having shifted their body clocks, they felt far less sleepy during the day and were happier with their lives. Their depression and stress scores improved, as did their performance in cognitive tests. They even got physically stronger.

I'm pleased to say that Marie followed this advice and now finds going to sleep and getting up at a normal time much easier.

Camping, Anyone?

An even quicker way to convert yourself from an owl to a lark could be to go camping. A few years ago, Dr. Kenneth Wright of the University of Colorado Boulder sent eight people (six men and two women) on a camping trip to the Rocky Mountains.[12] He gave them wrist monitors to record how much light they were being exposed to, and activity monitors to measure how much sleep they were getting. During the week that they were away camping, they were not allowed flashlights or cell phones, and the only light they saw at night was from candles or the campfire.

Their monitors showed that over the course of the week

they were exposed to four times their normal levels of light. This had a big effect on their sleeping patterns.

Before they went camping, their average bedtime was 12:30 a.m.; by the time they got home it was more like 11 p.m. Their sleep was now much more in sync with sunrise and sunset.

When he tested their blood in the lab, Dr. Wright found that the participants' bodies had started to release melatonin two hours earlier than they had before the trip. He had turned owls into larks in just one week. Who knows, it might even work on teenagers.

What Keeps Us Awake?

My idea of a great night's sleep is to go to bed around 11 p.m., fall asleep within a few minutes, and then wake up, refreshed, at about 7 a.m., without the need for an alarm clock. That would be lovely. It is what used to happen. It hardly ever happens now.

I have no problem getting to bed and falling asleep, but I almost always wake up in the middle of the night, and sometimes find it hard to get back to sleep. In this respect, I am a classic insomniac.

There are other types of insomnia: not being able to get to sleep is quite common, as is waking up early in the morning. But the most common form is waking in the middle of the night, particularly as we get older. This is partly because our sleep gets lighter as we age, but also because of things like having a full bladder and feeling the need to go to the toilet.

I used to get quite worked up about this. I resented the fact that no matter how tired I was, I'd wake up 4.5 hours after going to sleep (normally around 3:30 a.m.). I would go to the loo, get back into bed, and then just lie there for what felt like hours, worrying about not being able to get back to sleep, and worrying about how tired I would feel in the morning. Finally, I'd drift off, only to be dragged awake again by the alarm clock at 7 a.m.

Then, a few years ago, while researching a documentary about life in Victorian slums, I interviewed Roger Ekirch, a professor of history at Virginia Tech. He told me that my pattern—falling asleep, waking for a while, then falling asleep again—was how many people slept in preindustrial times. Apparently, people would go to bed around 9 p.m., sleep for about five hours, then get up at about 2 a.m. They would do household chores, visit friends, or "enjoy a bit of intimacy" before heading to bed again for a "second sleep."

Professor Ekirch believes that the pressures of the industrial age and the arrival of electric lights changed all that; sleeping continuously became the new normal. And, as the practice of sleeping continuously became more widespread, the idea of a "first" and "second" sleep faded from public consciousness. Even napping, which can be hugely beneficial (see page 106) and which used to be very common in hot countries, has largely been abandoned.

To support his claims that biphasic sleeping (sleeping in two blocks) has deep roots, Professor Ekirch pointed me toward research done by Dr. Thomas Wehr, a psychiatrist at the National Institute of Mental Health.[13]

In the early 1990s, Dr. Wehr conducted an experiment in which he persuaded a group of healthy volunteers to spend a month in a lab, where it was pitch black for 14 hours of the day.

By the end of the experiment, the volunteers were sleeping an average of eight hours a night, but not in one block. Instead, they slept for three to five hours, woke for an hour or two, and then fell back asleep for a second block of three to five hours.

Carol Worthman, an anthropologist at Emory University in Atlanta, thinks there may be something in Professor Ekirch's claims. She has studied the sleep patterns of hunter-gatherers who follow a preindustrial way of life. She says that interrupted or polyphasic sleep is quite normal among them. In many of the tribes she has studied, she has found about one in four people are up and active at any given point in the night. She thinks that there may be an evolutionary advantage to this, because when our remote ancestors lived out in the open, it would have been important that at least some of the tribe were awake, alert, and looking out for predators.

If, like me, you often find yourself awake in the middle of the night, you can console yourself with the thought that humans have probably been doing this for thousands of years.

Buoyed by these discoveries, I decided that rather than fight my "old-fashioned" sleeping patterns, I'd work with them. So these days I accept that I will probably wake at about 3 a.m. and plan accordingly. If I have an early start, then I aim to be in bed by 10:30 p.m. This gives me a roughly four-and-a-half-hour "first sleep."

When I wake around 3 a.m., rather than lie there fretting, I get up and go to another room, where I listen to music, meditate, or read a really boring book. I keep a special collection of books for this purpose. When I start to feel sleepy, which is normally after around 40 minutes, I go back to bed for three or so hours of "second" sleep.

Between my first and second sleep, I take care to avoid doing anything exciting or stimulating. If you are awake in the night, your goal should be to bore your brain into going back to sleep.

Since I have, slightly reluctantly, accepted that I am unlikely to return to sleeping for a whole night without a break, I've felt more rested, less stressed, and much less likely to nod off during the day. Try it for yourself, and let me know how you get on.

Snoring

Along with having a full bladder, one of the main reasons why people sleep badly is that they or their partner snore. I come from a long line of snorers. My father used to snore really, really loudly, like someone sawing logs. It was loud enough to be heard on the other side of the house.

I also used to snore at an incredible volume; in fact, my wife said that when we lived in London, I snored so loudly that I drowned out the sound of the metal beer barrels being delivered to the pub across the street first thing in the morning.

Although the caricature of a snorer is a fat, middle-aged man, women also snore. A few years ago, British newspapers outed a grandmother of four as "one of Britain's loudest snorers." She was recorded snoring at a window-rattling 112 decibels, which meant her snoring was louder than the noise made by a low-flying jet. According to the papers, it was loud enough to drown out a "diesel truck, farm tractor or speeding express train." Apparently, her husband coped by sleeping in the spare room and burying his head in a pillow.

I don't know if this was her problem, but the main reason most people snore is that they are overweight. If you are a woman with a neck size over 16 inches (41cm), or a man with a neck size over 17 inches (43cm), you are almost certainly a snorer.

As we get older and fatter, we snore more. That's because our throat gets narrower, our throat muscles get weaker, and our uvula, which is that finger-like bit of tissue that hangs down at the back of our throat, gets floppier. All these changes mean that when we breathe in, the air can't move freely through our nose and throat and into our lungs. Instead, the incoming air makes the surrounding tissues vibrate, which produces that horrendous snoring noise.

Snoring and Sleep Apnea

As well as being annoying, snoring can be a sign of obstructive sleep apnea (OSA), which is much more worrying. OSA occurs when muscles at the back of the throat relax and temporarily restrict or block airflow as you sleep, which leads to falling blood oxygen levels. This, combined with an increase in blood pressure, puts you at increased risk of having a heart attack.

It can kill you. The actress Carrie Fisher, famous as Princess Leia in *Star Wars*, died from a heart attack at the age of 60 while on a plane. The coroner said that the main contributory factors were untreated sleep apnea and a buildup of fatty tissue on the walls of her arteries.

An awful lot of people with sleep apnea go untreated because they think it is just snoring and that snoring is harmless.

I was on a train recently when an overweight middle-aged man named George introduced himself. He knew I wrote diet books but was keen to point out that he didn't believe in them and he certainly didn't need to diet.

I asked George if he had any trouble sleeping, and he admitted that he felt tired all the time. He also said he snored loudly, particularly after a few drinks, and that his wife had told him that there were times during the night when he stopped breathing.

He said he wasn't worried by this, but I was, particularly when he told me that he was a long-distance bus driver and kept himself awake with energy drinks. I suggested that he get his wife to stay up for an hour or two to count how many times he stopped breathing. "Stopping breathing" means not breathing for 10 seconds or longer. Other things to look out for in someone with sleep apnea are regular gasping, snorting, or choking noises, hypersomnia (excessive daytime sleepiness), and lack of interest in sex. If you have these warning signs, do discuss them with your doctor.

OSA affects about one in four men and one in ten women. Unfortunately, it is particularly common in bus and truck drivers, who tend to be overweight because they spend a lot of their working lives sitting on their bottoms eating junk. A recent study[14] of 905 Italian truck drivers found that about half suffered from a sleep-related breathing problem, making them dangerously prone to falling asleep at the wheel.

I pointed out to George that having untreated OSA doubled his risk of sudden death. The fact that he was starving his brain of oxygen every night also increased his risk of Alzheimer's and dementia. He looked pensive.

I explained that the best way to cure his snoring and his sleep apnea was to lose weight, fast. The reason I used to snore so loudly was because I used to have a 17-inch neck. When I put myself on the 5:2 diet, back in 2012, and lost 20lb (9kg), I also lost an inch of fat around my neck. I completely stopped snoring and our house was finally at peace.

I suggested to George that if he wanted to find out more about the advantages of rapid weight loss and how to do it safely, he should buy one of my books or visit thefast800.com. He said he would think about it. I like to think he followed through.

Rapid Weight Loss and Sleep Apnea

Although slim people can develop OSA, it is far more common in people who store fat around the neck. As I told George, if you are overweight, then rapid weight loss is the most effective way of curing snoring and OSA.

A study[15] carried out in Finland with overweight or obese patients diagnosed with mild OSA found that putting them on a rapid weight loss diet (800 calories a day for up to 12 weeks) cured more than half of them. They lost an average of 23.6lb (10.7kg), which dramatically improved their sleep as well as their hypertension, high cholesterol, and raised blood sugar levels.

The greater the weight loss, the greater the improvement. Ninety percent of those who lost more than 33lb (15kg) were cured of OSA, but even if they lost and kept off just 6.6lb (3kg), their chance of a cure was still 38%.

If you have OSA, but are not overweight, or are not motivated to lose weight, you might benefit from a CPAP machine. CPAP stands for continuous positive airway pressure. It is a machine that sits beside your bed and pumps air into a mask covering your nose and sometimes your mouth while you are asleep. The idea is that the pressure of the air keeps your throat open so you don't stop breathing.

It can be a lifesaver. A woman I worked with discovered she had sleep apnea and was prescribed a CPAP machine. Within a couple of weeks, she was transformed from a shattered wreck into a bundle of happy energy. "I had no idea how awful I felt in the mornings until I stopped feeling awful," she told me.

Because she was feeling so much more energetic, she decided to follow my advice, did the Fast 800 diet, lost 33lb (15kg), and was soon able to return her CPAP machine to the astonished sleep clinic.

You might expect the ferocity of snoring to gradually decrease as you lose weight. But this is very often not the case. There seems to be a critical tipping point. So don't be discouraged if, to start with, your snoring does not seem to be improving—get down to a healthy weight and it will get better. And remember that, even if it doesn't completely cure your snoring, you will get all those other health benefits of losing weight.

The only real alternative is a CPAP machine, which has significant downsides. There is the cost, the inconvenience of having to carry it around with you when you are traveling, and the fact that you have to wear a mask in bed every night. A bit of a passion killer, I would have thought.

Anti-Snoring Devices

In addition to CPAP machines, there are lots of anti-snoring devices out there, ranging from nasal strips that keep your nose open to "mandibular advancement devices," which push your lower jaw and tongue forward, with the aim of opening up your airway. Most of them work a bit, but none of them work as well as losing weight. If you get a referral to a sleep clinic, they may be able to offer you tailored advice; otherwise, your best bet is to go to Amazon and see what your fellow snorers have bought and what they make of their purchases.

If you are really desperate, there is also uvulopalatopharyngoplasty. This is an operation where a surgeon burns or cuts away tissue in your throat to try and clear the obstruction. This has risks, recovery is painful, and it is not always effective, particularly if the main reason you have sleep apnea is because you are overweight. It can make the condition worse.

Things That Go Bump in the Night

As you may have gathered by now, things can get quite eventful in the Mosley household at night. Not only do I regularly roam around at 3 a.m., but Clare sometimes gets up in the night, while remaining firmly asleep. Recently, she climbed over me and started looking through the clothes closet. When I asked her what she was doing, she said she was looking for a missing hamster that needed feeding. We haven't had hamsters for years. I persuaded her to go back to bed, and she immediately fell fast asleep and had no memory of any of it the following morning.

Clare has parasomnia, a common sleep disorder that includes a range of weird and wonderful things that can happen to people while they are asleep: sleepwalking, sleep talking, nightmares, sleep eating, sleep paralysis, sleep aggression, and even sexsomnia (having sex when you are asleep) are all forms of parasomnia.

About 10% of the population have a parasomnia of some sort, and though they can affect people at any age, they are more common in children, probably because children have immature brains.

Parasomnias also run in families, which could explain why two of my sons were sleepwalkers when they were young. We would often find them walking the corridors in the middle of the night. One of them managed to walk out of the front door, while fast asleep, and lock himself out. He spent half an hour banging on the door until one of us woke and let him in.

Sleepwalking can, of course, be extremely dangerous. When he was ten years old, our oldest son managed to sleepwalk his way out of the first-floor bedroom window of a cottage where we were staying. He fell 15 feet onto the flagstones below. We were unbelievably lucky that a neighbor, Russell, who happened to be wandering around outside at 3 a.m., heard him cry out, investigated, found him unconscious, and woke us. We wrapped his head in ice (I had recently made a documentary about how cooling can reduce the risk of brain injury) and he was rushed by ambulance to the local hospital, where an MRI scan revealed he had a fractured skull. To our huge relief he made a full recovery.

After that, we made sure that all the upstairs windows at home were secure. Fortunately, both boys stopped sleepwalking in their early teens.

Q&A

Should you try to wake someone who is sleepwalking, sleep eating, or having some other parasomnia episode?

It is not a good idea to shake or shout at a person who is in the grip of parasomnia, as that may trigger an irritable, aggressive, or even violent reaction. It is much better to mutter "time for bed" and gently guide them back to their bedroom.

At what time of night does it happen?

It can happen at any time of the night, but it typically occurs when the person is coming out of deep sleep. One part of their brain is still in deep sleep, so they are unconscious, but other parts of their brain are awake enough that they can walk, talk, even drive. I met a woman who used to drive to work during the middle of the night, waking up in the company parking lot with no idea of how she had got there. She solved this problem by locking her car keys in the safe every evening. Apparently, her unconscious mind could drive but couldn't remember the safe's combination.

Can you prevent parasomnia?

If your child or partner sleepwalks at the same time every night, you could try gently waking them about half an hour before they would normally sleepwalk. By doing this you disrupt their sleep cycle, and in some cases that is enough to stop their parasomnia. You will need to do this every evening for at least a week to break the cycle.

If they are at serious risk of hurting themselves or someone else, they should see their doctor—they may be offered CBT (cognitive behavioral therapy) or medication.

Summary

- The two main drivers of your wake-sleep cycle are adenosine (a chemical that puts you to sleep) and your circadian clock.
- Your primary circadian clock, the one in your head, follows a roughly 24-hour day. In some people it runs fast (larks); in others it runs slow (owls).
- The clock is reset every day by bright morning light.
- When children become teenagers they switch from being larks to becoming owls, which explains why so many prefer to stay up late and get up late. There are evolutionary reasons for this.

- Owls can become more larky by following a number of simple rules.
- One of the main reasons we sleep badly is because of snoring and sleep apnea. This is caused by too much fat around the neck, and the best cure is rapid weight loss.

3.

ARE YOU GETTING ENOUGH?

Sleep is essential for spring-cleaning our brains, putting our memories into long-term storage, and boosting creativity. Poor sleep, as we saw in Chapter 1, contributes to low mood, obesity, type 2 diabetes, and low sex drive.[16]

Far from being "a criminal waste of time," as the inventor Thomas Edison claimed, getting enough good-quality sleep is vital for our mental and physical well-being. But how do you know if you are getting enough?

According to the National Sleep Foundation, these are the targets you should be hitting at different stages of your life:

Age	Recommended hours' sleep
1–12 months	14–15 hours
1–3 years	12–14 hours
3–6 years	10–12 hours
7–12 years	10–11 hours
12–18 years	8–9 hours
18–65 years	7–9 hours
65+ years	7–8 hours

Most modern teenagers don't come close to hitting these targets, with fewer than half of British or American kids getting the required hours. Computer-crazed South Korean teenagers do even worse, with the average 17-year-old student managing just 5.7 hours a night.[17]

Adults get closer to their targets than teenagers, though according to a recent Gallup survey,[18] the average adult in the US still gets only 6.8 hours a night. That is a full hour less than adults said they were getting in 1942.

Surveys conducted in the UK come up with similar figures, with the average adult claiming to get 6.5 hours a night. Australians do slightly better, hitting an average of seven hours and 18 minutes.[19]

Bear in mind that "sleep" in this context really means "hours in bed." Since even good sleepers spend about 15% of their time in bed awake, if you are in bed for seven hours you are probably getting less than six hours' actual sleep a night.

The other thing about taking an "average" figure is that some people need more than average, while others will need less.

So how do you know if you are getting enough sleep? You could fill out a questionnaire, such as the Pittsburgh Sleep Quality Index, which you can find online, but I prefer the simplicity of the Sleep Onset Latency Test.

The Sleep Onset Latency Test or Spoon Test

The idea behind this test is to see how quickly you fall asleep during the day, if you are given a chance. Daytime sleepiness is a good measure of "sleep debt" and therefore of whether you are getting enough good-quality sleep at night. If you fall asleep while watching TV or at the cinema, then you probably have "sleep debt."

The great thing about this test is that it doesn't require any fancy lab equipment; you just need a metal spoon and a metal tray. The version I'm about to describe was developed by a famous sleep researcher, Professor Nathaniel Kleitman, from the University of Chicago.

On the weekend, or whenever is most convenient, you skip your usual morning coffee or tea. Then, in the early afternoon, anytime between 1 p.m. and 3 p.m., you go to your bedroom with a metal spoon and a metal tray.

You close the curtains, place the metal tray on the floor by your bed, check the time, then hang your arm over the side of the bed over the metal tray, clutching the spoon. Finally, you close your eyes and try to drift off.

The idea is that if you fall asleep, the spoon will drop from your fingers and hit the tray with a loud clang, waking you up. As soon as that happens, you check your watch to see how much time has passed.

- If you fall asleep within five minutes of closing your eyes, it means you are severely sleep deprived.
- Falling asleep within five to ten minutes is deemed to be "troublesome."
- Falling asleep after 10 to 15 minutes suggests you have a mild problem.
- If you stay awake for over 15 minutes, you're probably fine.

An alternative version of this test, which is more practical but less fun, is to go to bed in the afternoon, as described, but this time you just set an alarm on your phone to go off after 15 minutes. You then see if you drop off before the alarm goes off.

I told a friend, Sarah, about the Spoon Test and she decided to give it a go. Sarah normally goes to bed around midnight and wakes up at 5 a.m. She lies quietly for an hour or so, just listening to the sounds around her, and then gets up quite happily.

Because she had read so much about the importance of getting at least seven hours' sleep a night, she wondered if she was doing something wrong. Anyway, she did the Spoon Test and passed it with flying colors. As I said to her afterward, she must be one of those relatively rare people who can get by on less than five hours' sleep without having any obvious problems.

Her husband also did the test. He fell asleep within ten minutes.

The Multiple Sleep Latency Test

This is a more sophisticated version of the Spoon Test, normally carried out in a sleep lab.

When you arrive at the lab, you are attached to numerous machines (to record brain waves, eye movements, muscle tone, etc.) and asked to lie down in a dark, quiet room during the day. The scientists measure how quickly you fall asleep and how deeply. After 20 minutes you are woken up. Then, two hours later, you do it again. And then again. In fact, you do this a total of five times. This test is used to diagnose whether you have a sleep problem, and if so, what type. Do you have narcolepsy or idiopathic hypersomnolence? A breathing disorder or excessive daytime sleepiness? It is expensive, but it is the most reliable way of getting to the root of a persistent sleep problem.

Who Is the Most Sleep Deprived?

Teenagers

Most teenagers need between eight and ten hours in bed each night, but as we have seen, in many countries, including the US and the UK, less than half are getting the amount of sleep required by their growing bodies and brains. That is why I am a great believer in moving the school day later, to fit in with their owl-like tendencies. Lack of sleep limits teenagers' ability to concentrate and learn new facts, and it also leads to aggressive and risky behavior. One study[20] found that teens who average less than seven hours a night were twice as likely to have unsafe sex as those sleeping for longer.

Sleep-deprived teenagers also eat more junk food. This makes them overweight, anxious, and depressed, which makes their sleep even worse. Being sympathetic and helping your teenager improve their sleep hygiene (see Chapter 4) will make a big difference to family life.

Parents with Young Children

As anyone who has had kids will know, you spend the first few months after they are born wandering around like a zombie in the middle of the night, warming up milk and trying to stop them from crying. If that sounds bad, then listen to this: according to a recent study by Warwick University,[21] it takes new parents at least six years to get back to sleeping as well as they did before having children.

For this particular study, the researchers asked 4,659 would-be parents to keep a record of how well they slept, and then followed them for six years.

Moms lost an average of one hour's sleep a night for the first three months after giving birth, while dads struggled by with a loss of just 15 minutes' sleep a night.

Although things got slowly better, by the time their child was six years old, moms were still sleeping 20 minutes less than they had before pregnancy, while fathers were still having to make do with 15 minutes less. Both sexes were significantly less happy with the quality of their sleep than they had been before becoming parents.

First-time parents were the most affected, probably because they tend to be more conscientious. When our oldest son was born, we both leaped out of bed every time he whimpered. By the time our youngest child came on the scene, we were more inclined to let her settle herself, and she did.

The best thing to be said about this phase of life is "it will pass."

Older People

Sleep deprivation becomes a bigger problem as we get older, with more than half of those over the age of 65 saying they have difficulty getting a good night's sleep. It is a myth that older people need less sleep. They need just as much, but most don't get it.

Although you have more time on your hands when you retire, and one would imagine fewer responsibilities, as people get older they snore more and need to get up in the night more often to go to the bathroom. Older people also take more medication, which can interfere with sleep, and the quality of their microbiome (gut bacteria) tends to deteriorate. As we'll see, the microbiome can have a big impact on sleep (see Chapter 5).

Menopausal Women

Menopause and its aftermath can trigger severe insomnia. During menopause, levels of the hormones estrogen and progesterone fall, often leading to hot flashes, mood disorders, and sleep problems. About 60% of postmenopausal women report occasional insomnia. Snoring and OSA are also much more common after menopause.[22]

The Fast Asleep program is a very effective way of treating insomnia in postmenopausal women, but taking hormone replacement therapy (HRT) can also help. A randomized controlled trial[23] in which more than 400 postmenopausal women aged between 50 and 69 were given HRT or a placebo found that those taking HRT reported significantly less insomnia as well as fewer hot flashes, night sweats, and aching joints, and less vaginal dryness. You can get HRT in the form of pills, patches, gels, or creams.

Why Do We Need to Sleep as Much as We Do?

We know that not getting enough sleep has a big impact on your brain and your body. But why most of us need at least six to seven hours of sleep each night is more of a mystery.

Horses, giraffes, and elephants seem to get by quite happily on a couple of hours, while our fellow primates need considerably more than we do. Orangutans curl up in a bed in the fork of a tree and get a solid ten hours, snoring away sweetly like great, hairy, orange babies. Baboons, on the other hand, sleep on their bottoms while balancing on a branch high above the forest floor. They also sleep for about ten hours a night, though their sleep is rather more fragmented.

Some anthropologists think that the invention of the bed (or, more accurately, "a sleeping platform") by great apes, tens of millions of years ago, was a hugely important part of our evolutionary story. Sleeping platforms meant that, unlike the precarious baboons, our remote ancestors could sleep securely in the trees, safe from predators and blood-sucking insects. It also allowed them to get more deep and REM sleep, which presumably boosted their brainpower.

But if sleep is so important for brain development, why do humans, with the biggest brains of all the primates, sleep the least? The short answer is: nobody knows. One thing's for certain: the amount of sleep you need is not related to the size of your intellect. In our house we have a dog and a cat, and they spend at least half their lives sleeping. No one who's met either would describe them as particularly smart.

Jobs That Rob You of a Good Night's Sleep

Not many people think about sleep when they are choosing a career. But if you value your sleep, then don't join the police or become a firefighter. Taking a job working in transportation, communications, or construction is also likely to be a sleep killer. And if you decide to go into one of the caring professions, do bear in mind that doctors, nurses, paramedics, and care workers are all expected to work antisocial hours. What these jobs have in common is that they involve a lot of shift work, being awake when your body is desperate to be asleep.

During my medical training, I was frequently sleep deprived, but I never had to put in the sort of hours that my wife, Clare, did when she was a junior doctor. She was on a rotation where once a month, after a normal 60-hour work-week, she would embark on a weekend shift on a Friday night. She would then work through the weekend until Tuesday evening, before returning the next day to finish a normal week.

Like other doctors I've talked to, she sometimes got so tired that she experienced visual hallucinations. "I remember, at the end of one of these marathon weekends, walking down this long, old-fashioned Victorian corridor in the middle of the night and noticing that the whole corridor was snaking and twisting. I was finding it hard to walk straight and to anyone passing it must've looked as if I was drunk."

Falling asleep on the job is an occupational hazard if you are a junior doctor. As I mentioned in the introduction, Clare briefly fell asleep while assisting during an operation, although no one seemed to notice (they were probably as tired as she was).

A medical friend of ours, Philip, remembers doing a long weekend, after a busy week, and being called down in the middle of the night to see a patient in Accident and Emergency. He sat down on the bed to chat with the patient and the next thing he was aware of was the sound of his beeper going off. He had fallen asleep on the bed, in mid-conversation. The patient had left him a note, saying: "I've gone home, I'm feeling better and I thought you needed the bed more than I do."

It is hard to prove that sleep-deprived doctors and nurses are harming patients, though I don't think many of us would want to be operated on by a surgeon who had had less than four hours' sleep the night before. Unlike with pilots, there are no measures in place to ensure that medical staff have had enough sleep. Rather, there is tremendous pressure on junior staff to work far longer than they are paid to do, and well beyond what is legal. I have heard countless stories of junior doctors who are still working impossibly long hours simply because that is seen as the main way to get on.

How Long Can You Go without Any Sleep?

In 1983, Allan Rechtschaffen and colleagues from the University of Chicago reported the results of a gruesome sleep deprivation experiment they had performed on rats.[24]

They took eight rats and put them in a cage with a pool of water at the bottom. The rats could keep dry by standing on a metal disk. But as soon as they showed signs of falling asleep, the disk would rotate, so they had to scamper to stop falling into the water. Within a few days, the sleep-deprived rats had swollen paws, and were losing their balance and starting to lose weight. Within a few weeks, they were all dead. Yet when they were dissected, no obvious cause of death was found.

Sleep deprivation has been widely used as a form of torture but, as far as I know, no equivalent of the rat study has ever been done on humans. Thanks to an unfortunate Italian family, however, we have a good idea about what happens when humans are deprived of sleep for a long time. It isn't pretty.

The story begins in the early 1970s when an Italian physician, Dr. Ignazio Roiter, living in a small town in northern Italy, was asked by his wife, Elisabetta, to examine her aunt.

The unfortunate aunt had stopped sleeping and started to experience terrible hallucinations. They tried all sorts of drugs to put her to sleep, but nothing worked. Over the next few months she remained sleepless; her body wasted away, until eventually she died. No one could explain what had happened, so the family, though upset and puzzled, mourned and moved on.

Then another of Elisabetta's aunts began to show the same strange symptoms. She found it harder and harder to sleep until she stopped sleeping altogether. She eventually died the same horrible death.

Dr. Roiter began to investigate his wife's family tree and discovered that the same thing had happened to generations of Elisabetta's family, going back to the early 19th century.

When Elisabetta's uncle Silvano started to develop signs of severe insomnia, Dr. Roiter put him in contact with a sleep specialist, who was unable to do anything to help him. After Silvano died, his brain was sent to experts in the US, who finally discovered that Silvano, like so many other members of the family, had died from a disorder that was later called fatal familial insomnia (FFI). It is a genetic disease where the body starts to produce an abnormal protein, called a prion protein, which slowly attacks and destroys the thalamus, an area of the brain that is partly responsible for regulating sleep. The good news is that a genetic test is now available. The bad news is that there is no cure and, if you have the abnormal gene, there is nothing anyone can do to prevent you falling into a twilight world of perpetual insomnia that ultimately leads to insanity and death.

Going for Broke

Unlike Dr. Rechtschaffen's rats and families with FFI, there are some people crazy enough to voluntarily go without sleep for long, long periods of time.

The official record for staying awake was set in 1964 by a 17-year-old high-school student named Randy Gardner. He wanted to win the Greater San Diego School Science prize and he thought breaking the world record would be one way to do that.

In a recent interview,[25] he said the first two days were OK, but by day three he was moody and uncoordinated. By day five he had started to hallucinate. Tests showed he was having trouble concentrating and forming short-term memories. Despite this he kept going. Finally, after 11 days, he'd had enough. He was driven to a local hospital, where he was fitted with an EEG and then allowed to sleep.

He slept for 14 hours, for much of which he was in REM

sleep. He was kept in the hospital for a couple more days, then went back to school. His self-experiment won him the coveted science prize.

There were no obvious side effects at the time, but years later, when Randy was in his sixties, he developed severe insomnia, which he describes as "karmic payback." He still struggles with his sleep.

Although Randy holds the *official* record of 264 hours, others have gone longer. In 2007, Tony Wright, a Cornish gardener, managed to stay awake for 266 hours, while being filmed in a bar in Penzance. He passed the time by drinking tea, playing pool, and keeping a diary. His achievement is not in *Guinness World Records* because the publishers don't want to encourage others to try to beat it.

I met Tony a few years ago, when I was making a documentary about sleep. Among other things, I wanted to see how long I could stay awake and what effect it would have on me. I asked Tony to join in, to give me tips and moral support.

Before starting, we did some cognitive tests (test of memory, mood, reaction times, etc.) and then, after a decent night's sleep, we began what I think is the most challenging stunt I have ever attempted (and I've done quite a few).

The first 24 hours weren't too bad, though the tests we were doing showed that I was already falling apart. My reaction times had slowed right down, I had become intensely irritable, and I was hungry. When I did a car-driving simulation test, I kept on crashing.

Tony, however, was holding up remarkably well. He did fine on the car-driving simulation test and his reaction times, grip strength, and balance showed no signs of deteriorating. His mood was good and, if anything, he became more cheerful as the hours dragged by.

I struggled on, feeling more and more dreadful. I kept myself going by pacing around, singing, and playing pool. Neither of us was allowed caffeinated drinks or any form of stimulant.

There were peaks and troughs. We were in New York (the producer liked the idea of filming a program about not sleeping in the City That Never Sleeps), and I remember vividly, after 48 hours of being awake, the wonderful feeling of standing beside the river in Brooklyn and watching dawn rise over Manhattan. The rising sun touched something inside me. My mood lifted and I felt gloriously alive.

The reason for this little burst of euphoria was that while the pressure to sleep, coming from the buildup of adenosine in my brain, was still increasing steadily, the start of a new day meant that my circadian clock was now insisting that I wake up and get on with my life. For a few hours, these two systems battled away, with my circadian clock slightly on top, but by midafternoon I was feeling truly shattered and had to fight to keep my eyes open. We went off to play baseball and my reaction times were now so bad I missed the ball every time. Tony was in excellent form and smashed the ball all over the place.

When the sun set, I had such a feeling of dread that I knew I couldn't go on much longer. My blood pressure was up, my blood sugar had soared, I had a thundering headache, and I performed badly on all the tests, particularly anything involving concentration or memory.

Tony remained annoyingly cheerful, and did his best to rally me, but by around 11 p.m. I told him and the producer that I couldn't continue. I had begun to hallucinate that the hotel walls were caving in. I staggered up to my bed and took a final look at my watch. I had managed a miserly 64 hours without sleep.

As soon as my head hit the pillow, I was asleep. I slept for ten hours, and woke up feeling absolutely fine.

So why couldn't I go even three days without collapsing, while Tony had managed more than 11? He said he thought it was a combination of his diet and his training. He had put himself on what he called a "Stone Age" diet, consisting mainly of raw food, and he was convinced that this helped.

I wondered if he, rather like a dolphin, might have the ability to sleep on different sides of his brain, allowing him to appear partly awake while being partly asleep.

This is not completely crazy. A team from Boston recently found that on their first night in a sleep lab,[26] most people show signs on the EEG of being much more deeply asleep on the right side of the brain than the left. To see if the left side of the brain really was more alert than the right, the researchers made a noise. The EEG showed that the sleepers had detected the noise, but only on the left side of their brain. The right hemisphere continued in blissfully deep sleep.

It certainly makes sense, from an evolutionary perspective, that when you are sleeping in a new environment you might want to have part of your brain alert, in case there are predators around, while allowing the other half to have a good night's rest. But in humans it only seems to happen in the very short term. By the second night in the sleep lab, researchers found that both sides of the sleepers' brains were just as deeply asleep.

Sleep Mutants

I think a more likely explanation for people like Tony is that he is one of those exceptional individuals who are highly resistant to the pressures of sleep. They are rare, but they exist.

In August 2019, researchers from the University of California[27] announced that they had found a family with a genetic mutation that let them live quite happily on four hours' sleep a night. This family has a mutation in a gene called ADRB1, which alters activity in various regions of the brain that regulate sleep.

When the researchers bred rats with the same mutation, the rats slept for nearly an hour a day less than normal rats and it didn't seem to do them any harm. But the fact that short-sleep mutations are rare in humans suggests it might have a downside; otherwise, you would expect more of us to be able, like Tony, to get by on four hours a night.

A Few More Things That Happen When You Don't Get Enough Sleep

I've mentioned the effects of poor sleep on your memory, your mood, your weight, and your risk of getting dementia. But what worried me most about the fact that Clare was working insane hours as a junior doctor was that she was driving home from the hospital afterward.

Falling Asleep at the Wheel

In a recent survey[28] of more than a thousand British doctors, 40% admitted falling asleep at the wheel after a night shift, and more than one in four said they knew of a colleague who had died in a car accident after a night on call.

We have strict laws against drunk driving, but not against sleep-deprived driving. In the US, sleep-deprived drivers are responsible for an estimated 100,000 car accidents every year and over 1,500 deaths.

The impact that a bad night's sleep has on your chance of having an accident is dramatic. A recent investigation by the US Department of Transportation[29] found that drivers who'd had four hours' or less sleep the night before were 15 times more likely to be involved in a car crash than those who'd had at least seven hours.

Someone who is seriously sleep deprived is as dangerous as a drunk driver who is well over the legal limit.

In this study, just a little bit more sleep made a massive difference. Those who'd managed to get in five hours before hitting the road were "only" twice as likely to crash, while those with a good six hours under their belt were a mere 30% more likely to cause a pile-up than those who'd managed seven hours.

Many of us will, at some point, find ourselves falling asleep at the wheel. At least half the people I've quizzed admit to having done it. I do a lot of traveling and on more than one occasion I've been driving home, late at night, fighting to stay awake.

So what should you do if it happens to you?

The best thing would be to find a hotel for the night. If you can't find somewhere to stay, the next best thing is strong coffee and a quick nap. I usually stop at the nearest service station, buy a strong black coffee, and drink it. I then set my phone to wake me in 20 minutes, lie back in the car, and have a quick snooze.

The caffeine usually takes about 20 minutes to hit my brain, so when I'm woken by my alarm I am already buzzing. I don't drive straight off, but instead go for a stroll and make sure I really am awake before getting back in the car.

Studies have shown that a cup of black coffee, followed by a 20-minute snooze, will make you far more alert than either just having the coffee or just having the snooze.

The downside is you will probably arrive at your destination still buzzing with caffeine and find it hard to go to sleep. Better that than not arriving at all.

When the Clocks Change

Twice a year, every year, 1.5 billion people in 70 countries turn their clocks forward by an hour in the spring and back by an hour in the autumn. This translates into roughly 40 minutes' more, or less, sleep on the night of the change. The effects of this natural experiment are impressive.

First, changing the clocks alters your risk of having a heart attack. In a recent study,[30] researchers looked at the number of people admitted to hospitals throughout the state of Michigan with symptoms of a heart attack on the day after the clocks went forward or back. Moving the clocks forward led to a 24% increase in heart attack admissions, while moving them back resulted in a 21% reduction.

Secondly, you're more likely to have a car accident. A study published in the *New England Journal of Medicine* in 1996[31] found an 8% increase in accidents on the Monday following the clock change.

The clock change could also put you in jail. In a study with the wonderful title "Sleepy Punishers Are Harsh Punishers,"[32] researchers combed through the archives and found that judges in the US give defendants longer sentences the day after switching to daylight saving time compared to other days of the year. And what applies to grumpy judges applies to the rest of us. Deprive us of sleep and we become harsher and more judgmental.

Summary

- The amount of sleep you need varies hugely, depending on your genes and your age.
- Some adults can get by on less than five hours a night, but most of us can't.
- The most reliable way to tell if you are getting enough sleep is to lie down in a quiet room in the afternoon and see how quickly you fall asleep. If you fall asleep within ten minutes of closing your eyes, it suggests you're seriously sleep deprived.
- In addition to the long-term impact of sleep deprivation on your brain and body, one of the biggest risks you face after even a single bad night's sleep is having a car crash. Your judgment and reaction times deteriorate without you noticing.

4.

TRIED AND TESTED WAYS TO BOOST YOUR SLEEP

Getting a good night's sleep is crucial for mental and physical well-being, but how important is it to *you*? What are you prepared to do to get up feeling refreshed in the morning? How much are you prepared to change? Many of us are so used to being chronically tired that we've forgotten what it feels like to be fully rested.

The Spoon Test, or Sleep Onset Latency Test, which I described on page 59, is one good way of assessing whether you are getting enough sleep. I've come across lots of people who don't think they have an issue but who, nonetheless, fail this test when challenged.

Here's another good test to try, which doesn't involve making time to go to bed during the day. Answer "yes" or "no" to the following eight questions:

1. When your head hits the pillow, do you find it hard to go to sleep?
2. Do you wake up during the night and then find it hard to get back to sleep?
3. Do you wake up earlier than you want to and find it hard to get back to sleep?
4. When you wake up, do you feel exhausted?
5. Do you feel tired and irritable during the day?
6. Do you find it difficult to concentrate during the day because you're feeling tired?
7. Do you get mad cravings for carbs (cookies, cakes, or something sweet) during the day?
8. Do you nod off while watching TV, while in the cinema, or in a public place?

If you answer "yes" to three or more of these questions, it is likely that you have a significant sleep problem, e.g., insomnia. And, as I explained in the introduction, once you've developed insomnia, you are caught up in the cogs of a vicious cycle.

Insomnia changes your brain chemistry so that when you

go to bed, your brain is overactive and stops you from sleeping. Or you fall asleep but wake up in the night, usually to go to the bathroom—after which you struggle to resettle.

When you return to bed, you find it hard to get back to sleep because your sleep drive has reduced and your thoughts are now galloping around like a wild horse, ricocheting from one anxiety-inducing subject to another.

Although you will, eventually, drift off, when you wake in the morning you feel shattered. So you drink lots of coffee. But the trouble with coffee is that you develop a tolerance to it. So you drink more, or move on to something stronger (like an energy drink). This gets you through the day, but by the time you get home from work you feel exhausted. Despite your good intentions, you really don't want to take the dog for a walk or go to the gym. What you want to do is collapse on the sofa with a glass of wine.

After a late dinner, you nod off on the sofa while watching TV. You wake up, have a final glass of wine and a bit of cheese before heading for bed. But as you go up the stairs, you realize that you no longer feel quite so tired. The nap you had on the sofa has depleted your sleep drive. So when you get into bed, you spend the next half hour scrolling through social media or news click-bait.

Your partner falls asleep and starts snoring. You turn off the lights, but immediately start thinking about all the things you have to do tomorrow, and how important it is to get a good night's sleep. You begin to worry and fret . . .

Does this sound familiar? If so, then sleep restriction therapy (SRT) could be for you. I am going to explain how you do SRT in full in Chapter 6, as part of my Fast Asleep program. But the principle is simple: for a few weeks, you deliberately *reduce* the amount of time you spend in bed to ensure that when you do go to bed at night you are really tired. Which means you should fall asleep quickly, stay in deep sleep longer, and wake up less frequently during the night.

Critically, sleep restriction will break down the associations in your brain between "bed" and "bad sleep." These beliefs, whether they are conscious or unconscious, are playing a large part in stopping you from getting a good night's sleep. Think of it as a way of getting rid of a bad habit that you have fallen into.

Sleep restriction is the most radical and challenging part of the Fast Asleep program. The other key elements are:

- Eating to establish a "sleepy biome"
- Improving your sleep hygiene—your sleep environment (i.e., your bedroom) and your habits around sleep

If you've read my book *The Clever Gut Diet*, you will know just how fascinated I am by the microbiome, the trillions of microbes that live in our guts. We've known for some time how important these microbes are for controlling our immune system and our appetite. Now there is new and very compelling evidence that they also have a big role to play in regulating stress and how we sleep.

Among the thousand different species that live in their dark lair in the large intestine, there are "good" ones and "bad" ones. The good ones, also known as the "Old Friends," have evolved with us over hundreds of thousands of years and work hard to keep us healthy. Among other things, they manufacture 95% of the body's serotonin. Serotonin is known as the "feel-good hormone," because it contributes to our sense of well-being and happiness.

As well as affecting our mood, serotonin plays a role in regulating appetite, digestion, sleep, and sexual desire. Unfortunately, our guts also harbor plenty of "bad" microbes, microbes that create inflammation, contributing to anxiety, depression, weight gain . . . and sleepless nights.

The good news is you can swiftly change the mix of microbes that live in your gut by changing what you eat. In the next chapter, I will show you how.

Before that, though, we are going to look at other tried and tested ways to help you sleep better. I'm going to take you through a typical 24-hour day, showing you lots of things you can do to improve your "sleep hygiene," including how to establish good bedtime habits, how to define your optimum sleep window, and how to use mindfulness and breathing exercises to ward off the anxiety and negative thoughts that are the enemy of a good night's rest.

A note of caution: If you have mild sleep problems, then putting the following recommendations into practice will help. But if you have proper insomnia, they won't be enough. To properly reboot your brain and cure insomnia, you will probably also need to do a short course of sleep restriction therapy (see Chapter 6).

The good news is that once you've done SRT (and it should take no more than a few weeks) and cracked your insomnia, then you will find that eating gut-friendly foods and practicing the following sleep tips will stop you from falling back into a world of sleeplessness.

Establishing Good Bedtime Habits

The tricky thing about sleep is that it is such an individual thing. The amount, quality, and type of sleep needed vary not only from person to person, but also within each person's lifetime. No one size fits all.

That said, all the experts I've spoken to are passionate about the importance of treating sleep as a habit, one that you can improve with practice, and they all agree that getting into a regular wake-up-go-to-bed routine is a good starting point.

Your regular wake-up-go-to-bed routine is known as your "sleep window." I normally go to bed at 11 p.m., getting up at 7 a.m., and I aim to do this seven days a week. That's my sleep window. If you are an owl, you would, undoubtedly, be happier going to bed later and getting up later. But for people who have children, or who have to work normal hours, getting up much later than 7 a.m. isn't practical.

If you don't have young kids, and if you can persuade your boss to allow you to start your day a bit later, then do. You could try explaining that you can't help it, that you are a genetic owl and you will not only be happier but more productive if you are allowed a more flexible workweek. If that doesn't work, try converting yourself into a lark (see page 41).

For shift workers, setting a sleep window is also going to be tricky, but I will deal with that in Chapter 7. For the rest of us it should be doable.

The really hard part of having a sleep window is sticking to it, particularly on weekends. The temptation will be to sleep in, especially if you have had a late night. But if you are serious about wanting to sort out your sleep, then I'm afraid a lie-in is a temptation you are going to have to resist. The idea that you can catch up over the weekend on poor sleep during the week is a total myth.

Part of the problem with a long weekend lie-in is that you will mess up your circadian rhythms, which are so important for driving the urge to sleep.

If you normally go to bed at 11 p.m. during the week but decide to stay up till 2 a.m. on a Saturday night, and then get up three hours later on the Sunday morning, that will really throw your circadian clock out of sync. This is the social jet lag I talked about in Chapter 2.

The other problem is that if you get up on Sunday at 10 a.m., and then head to bed and try to sleep at 11 p.m., you may struggle because you will have a reduced sleep drive. You've been awake for three hours less than during the week, so there will be much less of an adenosine buildup in your brain urging you to go to sleep.

Obviously, there are nights when you will want to go out on the town, but the critical thing is that once you decide the time that you are going to get up in the morning, you stick to it.

Clearing Out the Bedroom Junk

One of my top tips for anyone who is about to go on a diet is, before they start, to clear their cupboards of tempting treats and unhealthy foods. If there are chips, cookies, or chocolate in our house, despite everything I know about how bad these foods are for my brain and my waistline, I will eat them. The best way to resist temptation is to avoid exposure to it.

The same is true of your bedroom. This should be for sleep and for sex, nothing else. If you have a TV in the bedroom, or you take your cell phone to bed with you, the temptation will be to use them, and that can be highly disruptive.

There's a widespread myth that light coming from your computer or cell phone (blue light) is bad for you because it switches off production of the sleep hormone melatonin. In reality, light levels produced by these devices are too low to do much damage and the real reason they are disruptive is because they are exciting your brain just when you want it to be nice and relaxed. I always put my cell phone on sleep mode and try to avoid looking at it in the hour leading up to bedtime.

Anyone who has kids will know just how disruptive social media can be. We had regular battles with our kids about not having laptops and cell phones in the bedroom when they were teenagers, but we felt it was a fight we had to win. Which we did. Mostly. Sleep deprivation is spectacularly bad for teen brains, and the whole point of social media is that it has been designed by evil geniuses with the sole purpose of keeping you hooked.

Again, it's like food. The reason we find foods like choco-late and chips so hard to resist is that they have been created and manufactured to be moreish. When I start on a bar of chocolate, I have to eat the whole thing. It's just as well I have never got to grips with Facebook or Instagram.

8 p.m.: Your Wind-Down Routine

If you go to bed with your head buzzing and your guts still trying to digest the snack you just ate, you will find it harder to go to sleep. A proper wind-down routine begins several hours before you go to bed.

Stop Eating

Ideally, you will have finished your last meal of the day at least three hours before you go to bed. That is what I was recently advised by Dr. Satchin Panda, a professor at the Salk Institute in the US and a world expert in chronobiology and circadian clock research. He is the scientist behind a form of intermittent

fasting called time restricted eating (TRE), which is practiced by numerous celebrities, such as Hugh Jackman and Miranda Kerr, as well as some of the major techies in Silicon Valley. The head of Twitter, Jack Dorsey, does it, as does Geoffrey Woo, the CEO of HVMN, a "human enhancement" company.

The idea behind TRE is that limiting the window within which you eat will help you lose weight, improve your cholesterol and blood sugar levels, make you sharper, and help you sleep better.

In one form of TRE, 16:8, you might stop eating at 8 p.m. and not eat again until midday the next day. It is called 16:8 because you go 16 hours without eating and eat all your meals within an 8-hour window.

Dr. Panda told me that he thinks most people will find 14:10 easier to stick to, and that is what he aims for himself. He has an evening meal with his family around 6 p.m. and then doesn't eat again until 8 a.m. the next morning.

Why does he think it is so important to stop eating several hours before bed? Well, it is mainly to do with your core body temperature. Your body temperature naturally starts to fall as bedtime approaches, driven by your circadian clock. This fall also helps trigger sleep. The trouble with late-night eating is it raises body temperature. When a late-night snack hits your stomach, your gut has to spring into action to break down and absorb the food you've just eaten. This increase in gut activity means your core body temperature will remain high, just when you want it to go down.

But I like to have a nice milky drink, or a bowl of cereal, just before going to bed!

Drinking cocoa or having a bowl of cereal just before bed may be soothing, but it's a bad idea. Your pancreas (which produces insulin) needs downtime and, driven by your circadian clock, will have closed down for the night. So it won't be ready for the big sugar hit that cereal or cocoa will deliver. This sugar hit will cause your blood sugars to rise and keep on rising into the night, which is bad for sleep as well as for your body. Any fat you eat in your snack will also cause levels of fat in your blood to rise farther and faster than they would earlier in the day and take longer to come down.

A few years ago, I did an experiment on myself in which I ate exactly the same high-fat, high-carb meal at 10 a.m. and 10 p.m. Whereas my fat and blood sugar levels rose and fell quite quickly after the morning meal, in the evening they were both still rising well after midnight.

Another reason to avoid a late-night snack or glass of milk is that any protein in the food will cause your stomach to release acid. If you suffer from acid reflux, stay away from anything but water in the two hours running up to bedtime.

One of the other major benefits of TRE is that it gives the lining of your gut, which takes a fearful bashing during the day, more time to repair itself. It's a bit like trying to repair a motorway; you can't do it while cars are driving up and down in the day, so you have to wait till nighttime to close it down.

If you don't give your gut time to repair, you may develop a condition called leaky gut syndrome, which occurs when bacteria that are living in your guts escape through the damaged stomach lining into your bloodstream, causing inflammation, bloating, and pain.

But what if you feel hungry? Dr. Panda told me that once you get used to his regimen you no longer get late-night munchies. In fact, you will find, after a couple of weeks, that if you do have a late-night snack it will leave you feeling uncomfortably bloated.

Alcohol: A Mixed Blessing

Alcohol is a tricky one. Almost everything I've read says that you shouldn't drink alcohol at night because, while it may help you drop off, it will lead to snoring and more fragmented sleep later on. Although that may be true for heavy drinkers, there is some evidence that points to potential benefits for light drinkers.

A couple of years ago, Israeli scientists recruited 224 teetotal diabetics[33] and randomly allocated them to drinking a medium-sized glass (5oz) of either red wine, white wine, or mineral water with their evening meal, every evening, for two years. The wine and water were provided free of charge and the empty bottles collected afterward to make sure they really were drinking regularly.

So what happened? Well, red-wine drinkers will be delighted to hear it was the group drinking red wine who came out on top. Not only were there significant improvements in their cholesterol and blood sugar levels, but they also reported better-quality sleep.

In another more recent study,[34] this time carried out by an American team, researchers found that exposing mice to small amounts of alcohol, equivalent to a human drinking a glass of wine, made their glymphatic systems (the channels in your brain that open up in deep sleep) more efficient at washing out the brain and removing waste.

What makes this study particularly intriguing is that the woman behind it, Dr. Maiken Nedergaard, is also the scientist who first revealed the existence of the glymphatic system in 2012. As she points out, "Studies have shown that while heavy drinking for many years leads to an increased risk of cognitive decline, low-to-moderate alcohol intake is associated with a lower risk of dementia. This study may help explain why."

I personally find that one glass of red wine with dinner has little effect on my sleep, but a couple of glasses makes it measurably worse. If you drink every night and suffer from insomnia, do try giving up drinking for a week and see what happens.

I recently met a woman at a party who said she used to drink half a bottle of wine every night and thought it was helping her sleep. But when she gave up drinking for a week (because she was taking antibiotics), she quickly noticed how much better she felt. "After years of waking up almost every night and fretting, now I almost always sleep through the night and I feel fantastic. I have so much energy. I still drink on special occasions but giving up routine drinking has changed my life."

9:30 p.m.: Find Something Soothing to Do

Dim the Lights

By 9:30 p.m., your pineal gland should be busy pumping out the hormone melatonin, which, in turn, will be orchestrating the rest of your brain, getting it lined up for a night of sleep. Melatonin levels typically start to rise about 9 p.m. and peak in the early hours of the morning.

Really bright light switches off the production of melatonin, particularly light in the blue frequency. That's why cell phone manufacturers now sell products that reduce the amount of blue light they emit at night. But, as I've said before, this is a bit of a con. The problem with cell phones and tablets is not so much the light they produce, but the fact that they stimulate the brain just when you need to slow down.

Ideally, you should switch off the bright lights in your house and go for more subdued lighting. Tim Peake, the British astronaut who spent six months on the International Space Station, told me they have recently altered the lighting on the space station so it gradually changes, over the course of a "day," from light that is more in the blue frequency first thing to light that is more in the red frequency as the "day" progresses, mimicking the light changes that happen back on Earth.

The Hadza, hunter-gatherers who live in Tanzania, shun artificial lighting and have no word for "insomnia." They gather around the campfire at night, to share stories and experiences, before heading to bed a few hours after the sun has set.

Have a Warm Bath

Instead of relaxing by knocking back a few drinks, you would be better off having a warm bath (with a few drops of an essential oil, such as lavender), reading a book, or listening to music. Studies have shown that having a warm bath or shower an hour before bedtime can help you fall asleep and stay asleep.[35]

But for it to have an effect, you probably have to be in the bath for at least ten minutes and it has to be at least an hour before bedtime. Why? Well, not surprisingly, getting into a hot bath will heat you up. It will also increase the circulation of blood to your skin, hands, and feet. When you get out of the bath, after a good, long soak, and don your kimono, your body will continue to radiate heat. In time, this will cool your core temperature down. The critical words are "in time." This whole cycle of heating and cooling takes about an hour. Simply dashing into the shower for a couple of minutes just before jumping into bed (which is what I tend to do most nights) isn't going to make you sleepy.

Listen to Music

According to the National Sleep Foundation,[36] studies have shown that older adults who listen to relaxing music before bed fall asleep faster, sleep longer, wake up less during the night, and rate their nights as more restful. Apparently, slow tunes with a rhythm of 60 to 80 beats per minute, which you're likely to find in classical, jazz, or folk music, are the most effective sleep inducers.

10:30 p.m.: Countdown to Sleep

Try Keeping a "To-Do" Journal

As well as keeping a sleep diary (to be filled in every morning), you might, as part of your final wind-down to sleep, try keeping a "to-do" journal. The idea is that you write down a list of the things you need to do the next day. This will, hopefully, mean you spend less time agonizing about your to-do list in the middle of the night.

Does this work? Yes, there is evidence that it does. A small study of American university[37] students found that those who spent five minutes writing about the day ahead went to sleep an average of nine minutes faster. Nine minutes doesn't sound like a lot, but it is similar to the impact of taking a sleeping pill. Keeping a journal also reduced the tendency to wake up in the night.

While you've got your journal out, you might also want to write in it three good things that happened to you that day. It can be anything from a friend admiring your clothes to watching a great sunset. Expressing gratitude, also known as "counting your blessings," is a proven way to reduce stress, one of the main causes of insomnia.

When I was young, I was quite religious and I used to kneel by my bed most evenings and pray. I would thank God for the good things that had happened and ask him to forgive me for the stupid and thoughtless things I had done. It was a good way of finding peace before getting into bed. I'm no longer a believer, but I find the practice of writing down three good things has a similar effect.

Thinking of, and then, importantly, writing down three good things works because it shifts your thoughts toward the pleasant things that happened during the day, helping to counter the natural tendency at night to ruminate and worry—it is these ruminations that frequently keep us awake.

Consider Taking Melatonin

There are dangers to taking sleeping pills, at least in the long term (for more information on sleeping pills and sleep remedies, see the Q&A on page 95). I take the occasional sleeping pill, such as zopiclone or zolpidem, when I travel long distance. It really helps me sleep on the plane and also combat jet lag when I land (see Chapter 7). I don't use it otherwise.

Melatonin is different. As I explained earlier, it is a hormone that is produced by the pineal gland, a pea-shaped structure in the middle of your brain. It is connected to your brain clock, the suprachiasmatic nucleus (SCN). When it gets dark, your SCN tells your pineal gland to start releasing melatonin. Rising levels of melatonin help coordinate the other parts of your brain that tip you into sleep. Levels peak at around 3 a.m. and then decline.

Synthetic melatonin is widely available and can be quite effective. So who should take it and when? As we get older, our brains tend to produce less melatonin, which could be why our sleep deteriorates. That is also why melatonin works best in people over the age of 55.

In the UK, Australia, and most of Europe, you can only get melatonin by prescription, whereas in the US you can pick it up at any pharmacy. I normally buy a few bottles when I am in the US, or order them online from a reputable American company like iherb. It is perfectly legal to buy melatonin this way, but it is illegal to sell it.

In the US, controlled-release melatonin is the recommended first-line treatment for older adults with insomnia. It has very few side effects (I have never noticed any); in fact, in one study[38] there were fewer side effects in those taking melatonin than in those taking placebo pills.

An Australian government report from 2011[39] concluded that taking 2mg of melatonin, one to two hours before bedtime, was safe and effective for people over the age of 55. They said it was safe to consume it daily for up to 13 weeks and that, unlike sleeping pills, there was no evidence that you get rebound insomnia when you stop. Taking it every night for months on end may not be a great idea because in a study in which people took melatonin or a placebo every day for six months, the researchers found that by six months there was not much difference between the two groups.[40]

Though it appears to be safe, the Australian government

report suggested it should not be used by children, pregnant women, or those with liver problems.

Although you are supposed to take it an hour or so before going to bed, I prefer taking it at 3 a.m., when I wake up and find it hard to get back to sleep. Since melatonin has a half-life of 3 to 4.5 hours, this should make me dopey the next morning, but it doesn't.

I take 2mg, slow release, but if that doesn't work you could try experimenting with higher doses (up to 5mg is safe). Unless I'm jet-lagged, I take melatonin no more than once or twice a week.

Q&A

Why can't I just take some sleeping pills?

By sleeping pills, most people mean benzodiazepines, such as temazepam, and the Z drugs, such as zopiclone or zolpidem. These can be helpful in the very short term for treating acute anxiety or distress of the sort that follows a bereavement or job loss. And I use zopiclone to help with jet lag. But like all medications, they have potential side effects and they become less effective over time.

If I start taking them, will I become addicted?

Modern sleeping pills are less addictive than the old-fashioned barbiturates, but they can become habit-forming quite rapidly. That is why a GP will rarely give repeat prescriptions, saving them for short-term use. If a patient is going through a more prolonged period of stress, affecting their sleep, then a GP might prescribe a low dose of amitriptyline—a drug used at higher doses for anxiety and depression, but which can improve sleep at a lower dose. It leaves some people feeling a bit dopey or "hungover" first thing in the morning, and it can have irritating side effects such as a dry mouth. It can also interact with other medications.

What about over-the-counter remedies?

There are a whole range of over-the-counter supplements and herbal remedies that probably help a bit if you have a short-term problem, but they don't offer a long-term solution because they do not tackle the underlying causes of insomnia. If you are taking other medications, you should consult your doctor as many of them will interact with other drugs. Also, be very careful what you take if you are pregnant or breast-feeding.

Chlorphenamine, also known by the brand name Piriton, is a popular over-the-counter antihistamine used to treat hay fever and insect bites and sometimes as a sleep remedy as it makes you feel drowsy. In rare cases it can cause allergic responses and may interact with other drugs, particularly some antidepressants.

Diphenhydramine, another sleep-inducing antihistamine, is found in products such as Nytol. Like chlorphenamine, it seems to be reasonably effective and has similar potential side effects.

A few small studies suggest that taking magnesium supplements may help elderly people fall asleep faster. On the whole, you would be better off boosting your magnesium levels by eating foods that are rich in magnesium, such as

avocados, leafy green vegetables, legumes, and nuts, like cashews and almonds.

Tryptophan can be taken in moderate doses as a supplement to improve sleep, but it has been found in some cases
to interact with other medicines such as antidepressants. It
can also cause side effects such as sweating, anxiety, nausea,
and vomiting.

Valerian is a commonly used sleep-promoting herbal supplement. There have been studies in which people taking
300 to 900mg of valerian before bedtime found it improved
self-rated sleep quality, but I could find no long-term studies. Again, watch out for drug interactions.

What about essential oils?

The most popular ones for helping you drift off are lavender,
vanilla, rose, and bergamot (a type of orange). You can put a
few drops in a warm bath, or you may prefer to scent the air
with a diffuser or spray. If you want to make your own, add
four to five drops of essential oil to half a cup of water and
then pour it into an empty scent bottle. You can also put a
couple of drops on the underside of a pillow or on the sheet,
but do not apply directly to the skin as these oils are highly
concentrated and can cause irritation.

Do they make a difference? I found one review article[41]
that concluded that they might help with mild sleep distur-
bances.

Lavender is the most widely studied. At the very least, it
will make your bedroom smell delightful and refocus your
thoughts on experiencing the lovely scent rather than on
your worries.

The Long Night of the Soul

Let's assume that you follow my advice and go to bed at 11 p.m., after having a lovely pre-bed routine, then quickly fall asleep and wake up feeling refreshed and thinking, "I must recommend this book to other people."

But what if you don't? What if you are still lying there, staring at the ceiling, listening to your partner snoring away, worrying about not going to sleep and besieged by negative thoughts?

You could try doing some breathing exercises or using cognitive behavioral techniques to help you challenge your thoughts (see page 100). On the other hand, it could be that when you went to bed you just weren't sleepy enough.

The rule of thumb is that if you feel like you haven't fallen asleep within 20 minutes of closing your eyes, then you should get out of bed and out of your bedroom. (I say "feel like" because I don't want you to be constantly checking the clock. It is better to just guess how much time has gone by.)

The technical term for this process is "stimulus control," and the reason you need to get up rather than just lie there is that you must associate bed with sleep and sex, nothing else. If you lie awake in bed, night after night, wrestling with your mental demons, this will trigger all sorts of unhelpful associations in your brain and body.

We have known, ever since the Russian physiologist Ivan Pavlov first demonstrated that he could make dogs salivate at the sound of a bell, how easy it is to create powerful, unconscious links. You do not want to reinforce, consciously or unconsciously, any associations between "going to bed" and "not being able to fall asleep." So if you can't go to sleep within what feels like 20 minutes, get out of your nice warm bed and head for another room, where you will sit and while away the time until you start to feel sleepy. This is not an excuse to go back on the laptop, catch up with a rerun of *Friends*, or check out Facebook.

If you are going to try melatonin, this may be a good time to do it. It will take about 30 minutes to kick in.

Ideally, you'll spend this time listening to soothing music or a dull podcast, or reading a book you have read before. I never have any difficulty falling asleep, but I often wake up in the middle of the night. Sometimes I am able to go back to sleep, sometimes not. I have a pile of books downstairs that I have read many times and that I work my way through as I wait for my sleep drive to reassert itself.

Night-Time Breathing Exercises to Reduce Stress and Pain

Yogic breathing (pranayama) is a form of controlled breathing that has been practiced for thousands of years. It is often done in conjunction with meditation or yoga. It reduces stress by activating the parasympathetic system, part of your autonomic nervous system. Activating the parasympathetic system causes your heart to slow and your blood pressure to drop. When I am struggling to sleep, the first thing I attempt is deep breathing, also known as belly breathing. In everyday life we tend to take shallow breaths, so this may feel a bit weird to start with.

You start by taking a slow, deep inhale through the nose, allowing the air to really fill your lungs. Put a hand on your belly—you should feel it inflate. Hold it for a count of two, then breathe out slowly through your mouth. The first few times you do it, it will feel unnatural, so you need to practice during the daytime. You will notice that as you do this, your heart rate will slow and you will start to feel more relaxed.

There are lots of different breathing techniques. The one I favor is 4-2-4.

- Breathe in deeply through your nose while mentally counting to 4.
- Hold your breath to a count of 2.
- Breathe out through your mouth to a count of 4.
- Try doing this for a couple of minutes. It should feel really relaxing.

You'll find more on breathing techniques on page 157.

Challenging Your Thoughts

Many people report that they are kept awake by crazy thoughts that seem to surface in the middle of the night. These could be anything from worrying about their kids to fretting about the damage that not sleeping is having on their brain. One way of dealing with these thoughts is to get out of bed and do something distracting, rather than engage with them.

But you could also try a bit of cognitive behavioral therapy (CBT). It is an approach that teaches you to challenge those thoughts. Although they are generated inside your brain, your thoughts are not real and they can be challenged in the same way you might argue with your dad when he says something particularly annoying.

Examples of the sort of thoughts you can be taught to challenge include:

"I won't get to sleep and if I don't, then I will feel really tired tomorrow and won't be able to function."

Challenge: "I'm sure I will get to sleep; I normally do. But even if I don't, it will be fine. One bad night's sleep is OK."

"It's the same every single night: I just lie here worrying. Why can't I ever get a good night's sleep?"

Challenge: "That isn't true: it doesn't happen every night. Yes, it's annoying, but it will pass."

Since negative thoughts like these are often part of a life-long pattern of thinking, successfully challenging them is rarely easy. That's why, if you want to learn how to do it effectively, you may need to see a qualified therapist.

3:30 a.m. and Wide Awake?

Although some people find it hard to go to sleep, the most common sleep problem is waking up in the middle of the night. It is often because you need to go to the bathroom, and once you have woken up it is hard to get back to sleep.

It typically happens four or five hours after you have fallen asleep, when you have been through three 90-minute sleep cycles and are in the light stage of the sleep cycle. To use my diving analogy, it's as if you have come to the surface for a swift gulp of air but find you can't then get back down to the deep again.

What should you do? Essentially, my advice is do the same as you do when you can't get to sleep in the first place: try the breathing, try challenging your negative thoughts, and if, within 20 minutes, none of that is working, get out of bed.

The main thing is to try not to worry about the fact that you are awake when you would much rather be asleep, or what this means for the following day. Going down that road will just make things much, much worse.

I find it useful to reflect on the fact that waking up in the middle of the night is a natural thing and that it used to be seen as totally normal before we decided that we had to sleep through the night, as I explained on page 45.

Some people use this period of wakefulness as an opportunity to enjoy a bit of night life. I was recently contacted by someone who likes to go out and do some photography at 3 a.m., before heading back to bed and a second sleep. Others find 3 a.m. is a good time to write.

Charles Dickens would frequently wander the streets of London at 3 a.m., and even wrote a book about it, *Night Walks*. These strolls (which lasted up to five hours) gave him powerful insights into poverty, vice, and drunkenness, which he used in his novels. As dawn approached, Dickens would often go to a local railway station to watch the morning mail arrive.

"The station lamps would burst out ablaze," he wrote. "The porters would emerge, the cabs and trucks would rattle to their places, and, finally, the bell would strike up, and the train would come banging in, knowing that sunrise was not too far away."

He would then head home, to bed.

Extreme Larkiness

If you are an extreme lark, you might just choose to get up and stay up.

The actor Mark Wahlberg likes to rise at 2:30 a.m. He

starts his day with prayers, followed by breakfast and then a workout. I know this because he often posts on Instagram under the hashtag #4amclub. His daily schedule looks something like this:

2:30 a.m.	Wake up
2:45 a.m.	Prayers and then breakfast
3:40 a.m.	First workout of the day
1:00 p.m.	Lunch
4:00 p.m.	Second workout of the day
5:00 p.m.	Shower
5:30 p.m.	Dinner
7:30 p.m.	Off to bed

Tim Cook, head of Apple, the most valuable company in the world, is also an early riser.[42] He gets up at 3:45 a.m., does his emails for an hour, and then heads to the gym before putting in a full working day. However bad my insomnia has become, I have never been tempted to do this.

Time to Get Up and Embrace the Day

Assuming you are not a teenager or an extreme lark, then 7 a.m. is a typical time to get up. Although some overachievers get up much earlier, others like Elon Musk (up at 7 a.m.), Amazon founder Jeff Bezos (7 a.m. to 8 a.m.), and Mark Zuckerberg, head of Facebook (up by 8 a.m.), seem to have more normal sleep patterns.[43]

Do a Brief Workout

The one thing that all these top execs do is exercise as soon as they get up in the morning. I try to emulate them. When I roll out of bed, the first thing I do most mornings is a series of resistance exercises, which include push-ups and squats (see the Appendix on page 267). My routine normally takes less than five minutes. I have to do them straightaway or I know I will never get around to doing them.

Unlike aerobic exercise (running, swimming, walking), resistance exercise is particularly good at building and pre-serving muscles. Doing resistance exercises is also a great way of improving the quality of your sleep. A recent review ar-ticle in the journal *Sleep*[44] concluded that "resistance exercise improves all aspects of sleep, with the greatest benefit for sleep quality. . . . In addition to the sleep benefits, resistance exercise training improves anxiety and depression."

The great thing about my regimen is I don't need to go to the gym and it doesn't cost me anything. Plus, it is over really fast.

Let in the Light

When I have the time, I take our dog for a 30-minute walk before breakfast. Part of the reason for doing this is to get the exercise, but mainly it is to expose myself to lots of early morning light.

The amount of light you experience in your house, or in the car on the way to work, is only a fraction of what you get when you are outdoors, even when it is overcast and gloomy. The point of going out into the early morning light is that it will reset your internal clock, and let your body know that the day has begun.

During the long, dark winter, or if you really struggle with getting up in the morning, you might want to invest in a light box. A good one produces 10,000 lux (a measure of light intensity), which is similar to the levels you get outdoors on a bright spring morning. Light levels in your house or at the office are more likely to be a miserable 25 to 50 lux.

The great thing about light boxes is that you can work at your computer or read a book with one on a table beside you.

. . . .

Light Boxes and SAD

I first came across light boxes when I was making a film about SAD (seasonal affective disorder), also known as the winter blues. People who suffer from SAD start feeling uncharacteristically low at the start of winter and perk up again in the spring. Many of us experience the winter blues to some degree, but about 5% of the population find the winter months truly crippling. Typical symptoms include low mood, feeling sleepy during the day (despite lots of sleep), and having a mad craving for carbs. To treat SAD, you need to use a light box for at least 30 minutes every morning.

A light box is also a good way to help turn an owl into a lark because it can help reset your internal clock. But, a warning: When you use your light box depends on what your sleep problem is. If you are a super-lark, waking up much earlier than you want (something that is common as people get older) and you are struggling to stay awake at night, then you should not use a light box in the morning.

In fact, you might want to shun the early morning light as much as possible. Instead, you should aim to get a good blast of light in the late afternoon, thereby delaying the release of melatonin.

. . . .

What about Breakfast?

As we all know, Breakfast Is the Most Important Meal of the Day. What you may not know is that the slogan first appeared in 1917 in a magazine called *Good Health*—a magazine that, surprise, surprise, was edited by none other than Dr. John Harvey Kellogg, one of the creators of Kellogg's Corn Flakes.

Although it is a slogan that has been endlessly repeated, there is no scientific consensus about the value of eating breakfast first thing in the morning. A recent meta-analysis published in the *British Medical Journal*,[45] which looked at evidence from 13 breakfast studies, concluded that telling adults that eating breakfast will help them lose weight is unwise because "it could have the opposite effect."

One advantage of breaking your fast later in the day is that it extends your overnight fast, giving your body more time to get on with essential repairs.

Is It OK to Have an Afternoon Nap?

Naps used to be a very Mediterranean thing to do. When I traveled around Greece and Spain in the 1970s, people would often head home in the early afternoon for a brief siesta. That way of life has largely gone.

The siesta was rebranded as the "power nap" in the early 2000s by James Maas, a former Cornell psychology professor turned corporate sleep guru. He claimed that a 15- to 20-minute snooze in the early afternoon was all you needed to recharge your body and brain.

There is something to this, though when and how long you nap is critical. Ideally, you should take your nap seven hours after you wake up (so if you woke at 6:30 a.m. your nap should be around 1:30 p.m.).

Your nap needs to be long enough to be refreshing, but not so long that you sink into a deep sleep. If that happens, you run the risk of sleep inertia: not only waking up feeling groggy and even drowsier than you did before, but also finding it harder to go to sleep that evening.

Though you can nap in a chair, it would be more effective to nap on a bed in a nice quiet room while wearing a sleep mask to block out the lights. Some employers, like Google, provide sleep pods for employees. Shift workers benefit most from having naps, and I will discuss that further in Chapter 7. Do remember to set your alarm because you really don't want to sleep the afternoon away.

Finally, Dinner...

And so we come full circle. If you want to try TRE, you might want to nudge your evening meal earlier so you finish eating by 8 p.m. at the latest. That's what we aim to do in the Mosley household.

You may also want to try some of the delicious sleep-enhancing recipes that Clare has created—you'll find them at the end of the book. In the next chapter, I am going to explain why some foods are so much better at helping you sleep than others.

Summary

- If you want to improve your sleep, you must first create a sleep window, the time at which you plan to go to bed and get up each morning, and stick to it as rigidly as you can.

- Improve your sleep hygiene by fostering good bedtime habits; limiting caffeine and alcohol; clearing electronic gadgets from your bedroom; and ensuring it is cool, dark, and quiet.

- Unlike sleeping pills, melatonin seems to improve sleep quality without causing addiction. There is limited evidence for the effectiveness of magnesium, lavender, or other over-the-counter remedies.

- Breathing exercises are an excellent way of slowing your heart and distracting your thoughts.

- If you wake in the night and find it hard to get back to sleep, get out of bed and do not return until you feel sleepy.

- Invest in a light box or go for an early morning walk or run. Half an hour of outdoor light every morning will help reset your internal clock.

- If you want to take a nap, make sure you do so no later than 2 p.m. and for no longer than 20 minutes.

5.

EATING YOUR WAY TO A GOOD NIGHT'S SLEEP

As I mentioned in the previous chapter, a few years ago I wrote a book called *The Clever Gut Diet*, about the impact that the food we eat has on our microbiome, the trillions of microbes that live in our guts. The book included recipes and advice, shown to boost the good bacteria and minimize the bad. Improving your microbiome can help you lose weight, boost your immune system, and improve your mood.

Since I wrote that book, there has been lots more research

into the impact of food on mood, including some that has looked at the effect that specific foods have on our sleep. For example:

- An experiment carried out by researchers from the Institute of Human Nutrition in New York[46] showed for the first time that feeding volunteers a diet rich in fiber and protein led to more deep sleep, while getting the same volunteers to eat foods rich in sugar and simple carbs led to more fragmented sleep.
- Another study, from the same group,[47] which looked at the diet and sleeping patterns of more than 2,200 people across America, found that those with a high M score (see page 113) slept longer and better than those eating a more typical American diet.
- A study published in 2017,[48] carried out by Professor Felice Jacka, director of the Food & Mood Centre in Melbourne, Australia, showed for the first time that putting people with moderate to severe depression on a Med-style diet led to such big improvements in mood that many were no longer clinically depressed.
- In a more recent study, published in October 2019,[49] researchers from the US showed that people who have higher levels of certain microbes in their guts, particularly those from the phylum *Bacteroidetes*, enjoy deeper, more efficient sleep and less waking at night than those with lower levels.

I will go into these and many other studies in more detail later on in this chapter, but first let's lay to rest some commonly accepted ideas about sleep-friendly foods.

A Load of Old Turkey

The idea that certain foods can aid sleep is not a new one. But you might be surprised to learn which really do impact sleep. They're not the ones that are commonly talked about.

There is, for example, a widespread myth that eating turkey makes you sleepy. According to a self-styled sleep expert I found on the internet, "Turkey is rich in an amino acid called tryptophan, which goes to your brain and is converted to serotonin, which helps you sleep."

In fact, turkey contains no more tryptophan than chicken or beef, and a lot less than nuts, seeds, or cheese. Another problem with this claim is that although eating lots of turkey will raise the levels of tryptophan in your blood, it has no effect on the levels in your brain because so little of it crosses the blood–brain barrier. For this reason, taking tryptophan capsules is also a waste of time.

Two other foods that I've seen hailed as sleep enhancers are tart cherries and kiwifruit.

Claims about the benefits of tart cherries are based on a couple of small studies done with elderly people with insomnia who drank a large glass of cherry juice (8oz) twice a day for two weeks.[50] There were some modest changes in sleep quality, but the notion that it works because "tart cherries contain melatonin" is clearly nonsense. You would need to drink about 132 gallons (500 liters) a day to get a big enough dose of melatonin to make a difference.

And as for kiwifruit, well, there was one small study from Singapore[51] in which the researchers asked 24 volunteers to eat two kiwifruit an hour before bedtime, every night for four weeks. The study found that it made a small difference, but when I tried following this regimen it made my sleep, if anything, worse. It also put me off kiwifruit for a long time.

Another popular myth is that eating cheese gives you bad dreams. Eating anything that is rich in saturated fat just before you go to bed is likely to disrupt your sleep, but there's no evidence that cheese is worse than any other food, or that it triggers nightmares. The claim that cheese causes nightmares seems to have started with Charles Dickens's book *A Christmas Carol*, in which the central character, Ebenezer Scrooge, blames his disturbing visions on eating cheese. But when, in 2005, researchers from Surrey University asked 200 men to eat different cheeses every night for a week and record their dreams, they found no evidence that eating cheese induced nightmares, though it may have resulted in more vivid dreams.[52]

So Which Foods Do Improve the Quality of Your Sleep?

Let's start with my favorite way of eating, the Mediterranean diet, the traditional diet of the countries that border the Mediterranean Sea. I have been a huge fan ever since I discovered that it is not only super-tasty but also has a wide range of health benefits. Studies[53] have shown that eating a Mediterranean diet will:

- Cut your risk of having a heart attack or stroke by around 30%
- Cut your risk of developing type 2 diabetes by 50%
- Cut your risk of developing breast cancer by up to 70%

The traditional Mediterranean diet involves consuming lots of olive oil, nuts, oily fish, fruit, vegetables, and whole grains. You also eat reasonable amounts of full-fat yogurt and cheese, as well as a glass or two of red wine with the evening meal. There is not much room in this diet for cakes, cookies, or highly processed food. You can calculate how "Mediterranean" your current diet is with the following simple quiz.

What Is Your M Score?

Add a point for each "yes" answer. A score of 10 or higher is good.

1. Do you use olive oil as your main cooking fat and dressing?
2. Do you eat two or more servings of vegetables a day? (1 serving = 200g/7oz)
3. Do you eat two or more servings of fruit a day? (No points for sweet tropical fruits)
4. Do you eat less than one serving of processed meat a day? (1 serving = 100g/3.5oz)
5. Do you eat full-fat yogurt at least three times a week?
6. Do you eat three or more servings of legumes—e.g., peas, beans, lentils—a week? (1 serving = 150g/5.25oz)
7. Do you eat three or more servings of whole grains a week? (1 serving = 150g/5.25oz)
8. Do you eat oily fish or shellfish three or more times a week? (100–150g/3.5–5.25oz)

9. Do you eat sweet treats like cakes, cookies, etc., fewer than three times a week?
10. Do you eat a serving of nuts three or more times a week? (1 serving = 30g/1oz)
11. Do you cook with garlic, onions, and tomatoes at least three times a week?
12. Do you average about seven glasses of wine a week?
13. Do you sit at the table to eat at least twice a day?
14. Do you drink sweet, fizzy beverages less than once a week?

SCORE : [ ] /14

Notes:

- Potatoes do not count as a vegetable.
- Sweet tropical fruits include melon, grapes, pineapple, and bananas.

- Processed meat includes ham, bacon, sausages, and salami.
- Whole grains include quinoa, whole rye, and bulgur wheat.
- Nuts include walnuts, almonds, cashews, and peanuts and should be unsalted.
- Drinking much more than seven units of alcohol a week can be harmful.

The Mediterranean Diet and Sleep

Although most of the research looking at the Mediterranean diet has focused on its impact on reducing heart disease, cancer, dementia, and diabetes, in the last couple of years there have been a couple of big studies published in reputable journals looking at its effect on sleep.

In May 2019, for example, a study in Italy[54] looked at the link between what Italian adults are eating and how well they are sleeping. For the MEAL study,[55] researchers collected data from 1,314 men and women living in Catania, one of the largest cities on the island of Sicily.

The researchers took detailed records of what the participants ate and then used the results of their completed food questionnaires to divide them into four groups, based on their M score, ranging from low to high Mediterranean diet adherence.

The participants also filled out the Pittsburgh Sleep Quality Index, a more detailed version of the test I mentioned on page 59.

When the researchers compared what they ate with how well they slept, they found that those with a high M score were more than twice as likely to enjoy decent-quality sleep as those with a low M score. They not only slept longer, but they had a higher sleep efficiency and were less likely to have a disturbed night.

Interestingly, this was true only for those who were a healthy weight, or a bit overweight. The men and women who were obese (with a BMI over 30) were not protected from poor sleep by a healthy diet.

These findings were replicated by another big study[56] that looked at the links between diet and sleep in more than 2,000 middle-aged men and women in the US. Again, there was a clear link between participants' M score and how well they slept.

The problem with observational studies, like the two I've just described, is that you can never be entirely sure of the extent to which it is a good diet that leads to better sleep, or vice versa. As I pointed out earlier, when people are sleep deprived they tend to eat more junk food.

That's why I was pleased to come across a novel intervention study conducted at Cornell University in New York,[57] in which the researchers manipulated the subjects' diet and then saw what effect that had on their sleep.

For this study, they asked 26 adults—13 men and 13 women—to spend five nights in a sleep lab, wired up to machines so their sleep could be monitored in detail. During this time, the participants ate meals that contained varying amounts of fat, protein, carbs, fiber, and sugar.

It turned out that when they ate meals containing more saturated fat, carbs, and sugar, they had lighter, more disrupted sleep. But when they ate meals that were richer in protein and fiber, they got to sleep faster and spent more time in deep sleep.

There are a number of reasons why eating a Med diet helps sleep. These include the fact that:

1. The foods in this diet, like olive oil, oily fish, legumes, and vegetables, contain anti-inflammatory compounds, such as oleic acid, omega-3 fatty acids, and polyphenols. We know that inflammation leads to arthritis and other painful conditions that keep people awake at night. We also know that neuroinflammation (inflammation of the brain), which becomes more common as we get older, contributes to poor sleep and dementia.

2. Going on a Mediterranean diet boosts levels of the "good" bacteria in your gut—these, in turn, can produce powerful anti-inflammatory agents, as well as "feel-good" chemicals that reduce anxiety. Since one of the main reasons people stay awake at night is because they are ruminating and fretting, anything that improves mood is likely to be good for sleep.

The Impact of the Mediterranean Diet on Mood

I am a big fan of Professor Felice Jacka, the dynamic director of the Food & Mood Centre at Deakin University in Melbourne, Australia, and the pioneering studies that she and her colleagues are doing, showing how the foods we eat influence our brain, mood, and mental health.

I came across her work in 2017, when she published the results of the SMILES trial.[58] It was the first intervention study looking at whether putting people on a healthier diet could improve depression. It was truly groundbreaking, and because it was so revolutionary it was also hard to get off the ground.

In retrospect, it is astonishing that it took scientists such a long time to run an experiment like this one. Felice had initially wanted to recruit 180 people with moderate to severe depression for the SMILES trial, but after three long, hard years of trying, she and her team had only managed to recruit 67.

The problem with having modest numbers of people in a study is that it is much harder to prove that your intervention has made a difference. Felice realized that the impact of the diet would have to be pretty dramatic for any results to be deemed "significant."

Once they'd recruited their 67 subjects, her team randomly allocated 33 of them to a dietician who helped them start eating what they called a "ModiMed diet," while the other 34 got "social support."

Those allocated to the ModiMed diet were encouraged to eat more whole grains, vegetables, fruit, legumes, and unsalted nuts, as well as some eggs and dairy. They were also asked to eat three tablespoons of olive oil a day, and fish and chicken at least twice a week. Perhaps surprisingly, they were also advised to eat a moderate amount of lean red meat, such as beef or lamb, three or four times a week. This was because research conducted by Felice and others had shown a link between eating red meat and mood, possibly because of the iron and vitamin B12 content.[59]

The participants were also asked to eat fewer unhealthy

foods, such as sweets, refined cereals, fried food, fast food, processed meats, and sugary drinks.

After all that preparation—the struggle to recruit volunteers and the meticulous preparation of the diet—Felice told me she was worried that they wouldn't find anything, so she was thrilled when the final results were clearer than she had dared hope.

Just under a third of the people who were put on the ModiMed diet had such large improvements in mood that they were no longer classified as "depressed." This was four times better than those getting social support.

There was also a significant improvement in the anxiety scores among the ModiMed group. The fact that it was those who made the biggest changes in their diet who saw the greatest improvements strongly suggested that it was the change in diet that had made the difference.

As one of the participants, who had already tried talk therapies and medication without success, later told the professor: "The program was to me a last resort. With its success I am forever grateful."

Another man, who had suffered from bouts of severe depression, wrote to say that taking part in the study had not only led to big improvements in his mental health, but in his sleep as well.

All of which is impressive and heartwarming, particularly when you appreciate just how hard depression can be to treat with conventional therapies.

What is really encouraging is that other, bigger intervention trials, like the HELFIMED study, have come up with similar findings.[60]

You might be thinking, "That's fine if you have money,

but eating healthily is going to be much more expensive, and therefore not an option for people on a tight budget," but in fact eating healthily can be cheaper than eating badly. The researchers in the SMILES study did a detailed analysis and showed that the cost per person for the diet they are recommending came to $112 per week. This was far lower than the $138 per week that the participants were spending, on average, before the experiment.

The key to eating a Med diet cheaply is to go for canned and frozen foods, which are just as nutritious, as well as fruits and vegetables that are in season. Fiber-rich legumes such as lentils, beans, and chickpeas are a cheap and healthy addition to a meal.

So why did improving their diet have this effect on their mood? Felice thinks it could be because the Med diet reduces inflammation and oxidative stress (the diet contains lots of antioxidants that help mop up free radicals, which otherwise damage brain cells).

But it could also be the diet's effect on the microbiome, which Professor Jacka's team are intensely studying. For more information, I recommend her book *Brain Changer*.

The Microbiome, Stress, and Sleep

You have one to two pounds of microbes, the weight of a large bag of sugar, living in your gut. The 100 trillion microbes that live down there are known as the gut microbiome, and there are as many of them living in your large intestine as there are cells in your body. Which means that you are 50%

human and 50% microbe. The human–microbe ratio is so finely balanced that a scientist recently calculated that every time you have a poo (your feces is 75% dead bacteria), you briefly tip over into becoming more human.[61]

Your gut microbiome consists mainly of bacteria, but there are also fungi, viruses, and simple, primitive animals called protozoa down there. In fact, most of us have at least 1,000 different species of microbes in our guts, fighting, reproducing, and competing. Together they form a wonderfully complicated ecosystem that I like to think of as my gut garden, or internal rain forest.

On the whole, we have a good relationship with our microbiome, which is not surprising as we have evolved with it over millions of years. We provide the microbes with a home and they help keep us healthy. We used to think their job was pretty basic: to protect our gut from foreign invaders; to synthesize vitamins like vitamin K, which the body doesn't make; and to produce nasty smells while gobbling up the fiber that our bodies can't digest. Gut gases, better known as farts, are the result of microbe activity.

Now we know the microbiome also:

1. Influences our body weight by shaping our appetite and cravings, and deciding how much energy our body extracts from the food we eat. Can your microbiome make you fat? It certainly can.
2. Teaches our immune system how to behave. If you don't have the right sort of microbes in your gut, you are at much greater risk of a range of allergic and autoimmune diseases, from asthma to multiple sclerosis.
3. Last, but by no means least, it has a big impact on our mood and our sleep, so much so that the study of how the creatures in our guts affect our brains has its own name, "psychobiotics."

But how do tiny microbes, who live in your colon, at the far end of your gut, do all these things? Well, they may not have teeth or claws, let alone arms or legs, but they are brilliant chemists. Some of them can take the bits of food our body can't digest (such as fiber) and convert it into hormones that influence our mood—hormones like dopamine, serotonin, and GABA (a neurotransmitter that acts in a similar way to the anti-anxiety drug Valium).

Others can turn fiber into a chemical called butyrate, which is brilliant at damping down inflammation. Chronic inflammation is behind a whole range of diseases, including cancer and heart disease. As well as lowering inflammation, butyrate helps to maintain your gut lining, the barrier that keeps bacteria and other toxins from escaping into your blood. The recipes at the back of this book contain foods proven to boost your butyrate levels.

So one of the secrets to a long and healthy life is having a wide range of helpful microbes living in your gut. Like a rain forest, you want your microbiome to be a rich, flourishing population of different creatures.

It's All About Diversity

Greater diversity in the gut means having a much more capable and resilient microbial community, one in which there is more competition, so no one species can dominate. It also means hosting lots of different microbes with different chemical talents. We benefit from having firemen-microbes to put out inflammation, builder-microbes to repair the gut wall, and pharmo-microbes to make drugs that help us sleep better.

Conversely, we know that having a narrower, less diverse microbiome is associated with a range of chronic diseases, including a higher risk of obesity, inflammation, type 2 diabetes, colorectal cancer, and allergies.

Sadly, the microbiome tends to become less diverse as people grow older. This is partly because people tend to eat a narrower range of foods as they get older and partly because they eat more "convenience" foods, which are highly processed. One of the problems with ultra-processed foods is they often contain emulsifiers, which are added to extend the food's shelf life. Your microbes do not enjoy them.

By the time most people are in their sixties, healthy microbes, like *Lactobacillus* and *Bifidobacterium*, have begun losing ground to pro-inflammatory, opportunistic bacteria. They are called opportunistic because they can cause an infection when given half a chance. Older people also take more prescription drugs, which can play havoc with the microbiome, and they do less exercise, which again reduces microbiome diversity.

There's good evidence that people who manage to keep their gut garden in good shape as they age not only develop less chronic disease but also sleep more soundly.

If you are interested in finding out how diverse your gut garden is and which species are currently living in it, you can have your poo tested. There are lots of different companies, including British Gut and American Gut, that will sequence your microbiome and provide you with a report of the results for a relatively inexpensive rate.

The process is simple. You can pay for the service on their website and they will send you a plastic tube, a spatula, and instructions on how to collect a fecal sample. You put your feces in the tube, give it a good old shake, stick it in the mail, and wait for the results. And wait. You can expect the whole thing to take at least two months.

A while ago I sent my feces off to be tested, and the results were certainly interesting. Among other things, they told me that my Simpson index (a measure of just how diverse the biome is) had a score of 7.99. That puts me into the top 30% of "most diverse microbiomes" of those who have had poo samples tested. Good, though clearly room for improvement.

What Is the Link Between Microbial Diversity and Sleep?

In a study published in October 2019,[62] researchers from Nova Southeastern University in Florida asked a group of men to wear activity monitors and have their sleep patterns analyzed over the course of a month. The researchers also collected lots of poo samples from the men, which they then examined closely.

One of the things that came out of this research was clear evidence that the men with the most diverse microbiomes had the best-quality sleep, which included longer total sleep time, higher sleep efficiency, and much less nighttime waking.

As well as confirming the importance of diversity, researchers identified a number of gut species that were present in higher numbers in the men who slept well. These included *Bacteroidetes*, which produce GABA, a neurotransmitter that promotes sleep, and *Corynebacterium*, which makes another neurotransmitter, serotonin, which again has been shown to promote sleep.

Another interesting finding was that the men with greater microbiome diversity also had higher levels in their blood of interleukin-6, a cytokine that plays an important part in regulating the immune system and influences both sleep and memory.

In an experiment conducted in 2009,[63] German researchers asked a group of healthy young men to spend two nights in their sleep laboratory. Before going to bed each night they were asked to read a short story. Then, the researchers sprayed a fluid containing either interleukin-6 or a harmless placebo into their nostrils. In the morning, the men were asked to write down as many words as they could remember from the story they had read the night before. Inhaling the interleukin-6 clearly made a difference, because not only did they get more deep sleep, compared to the night when they inhaled a placebo, but they could also remember more words from the story.

What all this research suggests is that boosting the good bugs in your microbiome is beneficial for your body and brain, and also has a significant impact on the quality of your sleep. A win-win. So how do you do it? You could start by eating more prebiotics and probiotics.

Prebiotics

A prebiotic is a type of nondigestible plant fiber that acts like a fertilizer to encourage the growth of "good" bacteria in your gut. Although many vegetables are rich in fiber, they aren't all classified as prebiotics because many don't contain the sort that will really boost your microbiome. Here are some of the best prebiotics.

Beans and Lentils

Beans and lentils form a big part of the traditional Mediterranean diet. As well as being a great source of prebiotic fiber, they are rich in B vitamins, which have been shown to help sleep, and protein, making them a great alternative to meat. They can be used in soups and stews, with or without meat, and are delicious in curries. Hummus, which you can buy or make yourself from canned chickpeas, is a nutritious dip, eaten with veggie sticks. Beans and lentils are my top sleep-inducing foods. For ideas on how to enjoy more of them yourself, check out the recipe on page 219.

Onions, Leeks, and Garlic

These three members of the Allium family are packed with antioxidants and other nutrients, and are a good source of prebiotic inulin. I love this family of plants and cook with them on a regular basis. Our recipes are full of them. The Spanish make a wonderful tomato base called sofrito, which consists of garlic, onion, paprika, and tomatoes all cooked together in olive oil. It's delicious with chicken, fish, or shrimp.

Endive and Radicchio

Endive and radicchio, vegetables usually eaten in salads, also contain prebiotics. Endive root is particularly rich in inulin—it accounts for nearly half its fiber (see recipe on page 202). You may also encounter endive root as a caffeine-free coffee substitute.

Jerusalem Artichokes

Over 70% of the Jerusalem artichoke's fiber comes from inulin, making it one of the richest sources of this particular prebiotic. Some people call it "fartichoke" because the high levels of nondigestible carbohydrate often lead to flatulence. If you aren't a regular vegetable eater or you have IBS, Jerusalem artichokes are probably best avoided. Otherwise, you might like to try our recipes for Jerusalem Artichoke Soup (see page 194) and Beef and Jerusalem Artichoke Casserole (see page 228).

Whole Grains

Whole grains are also an important part of the traditional Mediterranean diet. Very low carbers may shun them, but whole grains are full of the sort of fiber your good bacteria will love. The trouble with the grains we most commonly eat, such as wheat and rice, is that by the time they get to us most of the fiber and other nutrients have been stripped out. Switch from white to brown rice, and try adding a few other whole grains to perk up your microbiome.

Oats

The most popular way to consume oats is in the form of oatmeal. Although I am a fan of oatmeal (we eat ours with added wheat bran and some toasted walnuts), I would urge you to avoid the highly processed instant stuff. You can cook proper oatmeal on the stove, or in the microwave, in a few minutes, or you might like to try our recipes on pages 187–188.

Barley

Barley is an ancient and very tasty grain, with a slightly nutty flavor. It is good in soups and stews. Barley scores well as a prebiotic because, like oats, it contains lots of beta-glucan. Beta-glucan is a soluble fiber that's been shown not just to boost your biome but to improve your cholesterol levels. It binds cholesterol in your gut, preventing it from being absorbed.

Flaxseeds

This is another healthy prebiotic, a seed with a slight, but not overwhelming, nutty flavor. You can scatter flaxseeds on your cereal in the morning, or add them, toasted, to salads for a delicious crunch. They are rich in insoluble fiber, so as well as feeding your microbiome, they should help ensure regular bowel movements.

Fruit

Apples and pears are high in microbiome-friendly fiber, particularly if you eat the skin. A medium-sized apple with its skin has 4 grams of fiber, while a medium-sized pear has 5 grams. I love stewed apples with yogurt, or diced and cooked in the oven with a scattering of cinnamon. We don't bother peeling them, even in a crumble.

Strawberries, blackberries, and raspberries are also surprisingly high in fiber and low in sugar. They're an excellent source of vitamin C and contain decent amounts of folate (vitamin B9).

Seaweed

Slippery, slimy, and with a very distinctive ocean flavor, seaweed is certainly an acquired taste. The seaweeds you are most likely to have encountered are nori, which is used to make sushi; dulse, which I was told tastes like bacon when fried—it doesn't; and kelp, which is used as a gluten-free alternative to noodles. Seaweeds are excellent prebiotics, packed with vitamins and minerals, as well as fiber, and probably the best source of omega-3 fatty acids. If only they didn't taste quite so strong . . .

Cocoa

I am a hopeless chocaholic and will eat bars of it given a chance. Although milk chocolate, which is stuffed with fat and sugar, is clearly bad for us, cocoa itself is surprisingly healthy. Unsweetened cocoa powder contains more than 30% fiber and it is an excellent source of flavonoids and polyphenols, both good for your gut bacteria.

Prebiotic Capsules

I am normally very skeptical about the benefits of swallowing vitamin pills, fish oil capsules, or supplements of any type. But a few years ago, I did agree, as part of a film I was making about sleep, to try out a product called Bimuno. This is a fiber supplement containing a prebiotic called galactooligosaccharide (GOS). There is good evidence that regular consumption will increase levels of the good bugs in your gut, such as *Bifidobacterium*. It is primarily taken to help with gut problems, but under the guidance of Professor Phil Burnet, a neuroscientist based at Oxford University who specializes in the effects of gut bacteria and prebiotics on brain function, I decided to see if it would have any impact on my sleep.

Phil provided me with a special sleep tracker, which I wore for a week (to determine my baseline). Then I consumed the Bimuno powder for a couple of weeks (it comes in a sachet and you stir it into a cup of tea or milk). Then I stopped taking it.

At the end of my little self-experiment, I sent the tracker back to Phil and a few days later met up with him to discuss my results.

Despite my skepticism, consuming the powder did seem to have made a difference. I noticed that I had begun sleeping better a few days after I started taking it, and then reverted to my more normal restlessness a few days after stopping. And that is also what the tracker showed.

As Phil explained, "If we look at the days before you took the supplement, 79% of your time in bed was spent sleeping and 21% of your time in bed was spent awake." In other words, my sleep efficiency was a lousy 79%.

"But five days after taking the supplement," Phil went on, "your sleep efficiency went up to 92%, which is an impressive turnaround. One thing that I was puzzled by was that on this night"—he pointed at my graph—"your sleep efficiency suddenly went down; do you know what you were doing?"

"I went out and had a few drinks," I admitted. "I wondered at the time whether you'd notice."

Is drinking a white powder the only way to get decent amounts of GOS? Phil suggested that I might get similar results from eating lentils, chickpeas, butter beans, lima beans, and cashew nuts, but because the powder contains higher doses, obtaining it through food might take longer to have an effect.

When the sleep film went out in the UK, there was a big rush to buy Bimuno products. In fact, I still, occasionally, get stopped in the street by people who say they saw the program and started taking Bimuno, and that it has transformed their sleep. Others, however, said it had made them windy, and it certainly can inflame the guts of people with IBS.

Probiotics

Just as important as feeding our microbiome with prebiotics is ensuring that we have the right balance of gut-friendly microbes in the first place. This is where probiotics come in. Probiotics are live bacteria or yeast that you parachute into your intestine, in the hope that they will take root and enrich your microbiome. There are lots of probiotic capsules and supplements out there, but as I've said before, I prefer topping up my "good bacteria" through food.

Yogurt

Yogurt is a key part of the Mediterranean diet and it is also a good source of the probiotic *Lactobacillus*. I like full-fat, plain Greek yogurt. I add fruits to sweeten it, or scatter cinnamon, flaxseeds, and nuts over it.

Cheese

Like milk and yogurt, I prefer my cheese full fat and unadulterated. Although I don't believe that eating cheese will give you nightmares, I would recommend you don't eat it close to bedtime. Not all cheese contains live bacteria, and processed cheese contains hardly any. You will find significant numbers of "good" bacteria in Gouda, mozzarella, cheddar, and cottage cheese, as well as in blue cheese such as Roquefort.

Fermented Foods

You can ferment almost anything, from vegetables to fish, but if you are not familiar with fermented foods, I would stick to some of the more standard ones, like sauerkraut or kimchi.

They are particularly delicious when they are homemade. Clare describes how to do this in the *Clever Gut* books, and there are a couple of new recipes in the recipe section on page 262.

Although having a more diverse microbiome seems to help sleep, there is not a lot of research out there showing whether probiotics themselves have an impact. I've seen one study from Japan[64] that found that women eating fermented foods during pregnancy were more likely to have babies who sleep soundly during their first year of life (the baby gets its microbiome from Mom), but unless you are used to them I would be cautious about including a lot of fermented foods when pregnant. Anecdotally, I've heard it helps some people but disrupts sleep in others.

Probiotic Capsules

As with probiotic foods, there is not a lot of research on the impact of probiotics on sleep. I found one small study, published in March 2019 by neuroscientists based at the University of Verona, in Italy.[65] They asked 38 male and female students to consume a probiotic capsule containing a mixture of four different bacteria (*Lactobacillus fermentum*, *L. rhamnosus*, *L. plantarum*, and *Bifidobacterium longum*) or a placebo, every day for six weeks. The students didn't know which they were getting.

The students filled in questionnaires that assessed their mood and how well they were sleeping at the beginning, middle, and end of the experiment.

The scientists found that, unlike those swallowing placebo capsules, the students who were taking the probiotics saw modest improvements in both mood and sleep quality.

· ·

Caution for those with IBS, gut problems, or reduced immunity

I mentioned earlier that for some people it takes time to boost a healthy microbiome, and if you are not used to eating fiber you may get some flatulence, bloating, and possibly a bit of diarrhea. This usually settles down—just build up your fiber intake slowly. However, those with IBS may need to proceed more cautiously or avoid adding fiber and fermented foods. If your immunity is suppressed, check with your doctor first.

· ·

Time Restricted Eating

In the last chapter, I suggested that you might want to try this very popular form of intermittent fasting by having your evening meal a bit earlier and your breakfast a bit later. Practicing TRE really can improve the quality of your shut-eye.

In a recent study run by scientists from the Salk Institute,[66] overweight volunteers who restricted their eating to a ten-hour window (14:10) lost an average of 7.3lb (3.3kg) over the course of 12 weeks. They also saw big reductions in waist size, blood sugar levels, blood pressure, and "bad cholesterol" levels. Last, and by no means least, most of them also enjoyed longer and less interrupted sleep.

Getting used to TRE can take time, but most people find that they soon adapt. I suggest you start by going 12 hours without eating (12:12) and then try to extend your overnight fast to 14 hours, i.e., 14:10.

Summary

Changing what you eat and when you eat can help you to sleep better:

- Try not to eat within three hours of bedtime.
- Experiment with extending your overnight fast to 12 and then to 14 hours.
- Cut back on sugar and sugary treats, drinks, and desserts, particularly shop-bought ones. They lead to more fragmented sleep.
- Get more fiber into your diet by switching to brown rice, and by eating more quinoa, bulgur, whole rye, whole grain barley, wild rice, buckwheat, lentils, and beans.
- Rolled oats are great for breakfast, as long as they are not the processed instant sort.
- Full-fat yogurt is a good source of probiotics. Add blackberries, strawberries, or blueberries for flavor. Or a sprinkling of nuts.
- Snack on nuts: they are a great source of protein and fiber, which should lead to more deep sleep. Try to avoid salted or sweetened nuts, which can be moreish.
- Eat oily fish such as salmon, tuna, and mackerel, which are rich in omega-3 fatty acids, two to three times a week. An experiment conducted with American prisoners found that eating oily fish can improve sleep.
- The best fruits to give your biome a treat are berries, apples, and pears.
- Have an alcoholic drink, if you must, but only with your evening meal. Try to average no more than a glass a day and stick to red wine, which is better for you than white.

6.

THE FAST ASLEEP PROGRAM

I've given you the latest science and identified some of the issues that prevent you from getting to sleep and staying asleep, along with advice on how to improve your sleep hygiene. I'm now going to pull all this information together into a four-week program. Follow it, and you will soon be feeling far more cheerful, with more energy and greater mental clarity. The first thing you have to decide is whether you want to try sleep restriction therapy (SRT), the part of the program that I introduced you to in Chapter 4.

SRT is extremely effective, but it is a challenge. Initially, during the day, you may feel sleepier and more irritable, so you do have to be very careful about driving or working machinery. Staying up late at night when you want to be asleep is also quite boring. The good news is that it doesn't last for long and it really does work.

Note: If you think you might have a significant health problem, or a sleep disorder such as sleep apnea, consult your doctor before starting on a new program such as this one. SRT should not be tried by pregnant women or young children.

Before You Start

As the American president Abraham Lincoln once said: "Give me six hours to chop down a tree and I will spend the first four sharpening the axe." The following is a list of things I recommend you do before you start the program.

Keep a Sleep Diary

On the next page is a sample page from a sleep diary. You can download and print a similar one from my website, fast-asleep. com. I want you to fill in your sleep diary for a week before you start the program and then every week during the course.

The point of keeping a sleep diary is to assess how well you are sleeping and to help you calculate your sleep efficiency, i.e., the amount of time you spend in bed actually asleep. Just to remind you, if you are in bed for eight hours but asleep for six hours, then you have an efficiency rating of 6/8 = 0.75 or 75%, which is poor.

By the end of the program, you should be aiming to get it up to 85%; 100% is unrealistic because everyone needs a bit of time to drift off. If you are falling asleep immediately when your head hits the pillow, I would be worried that you are too sleep deprived.

If you don't get as high as 85%, don't worry. An efficiency rating of 80% is perfectly acceptable, particularly for older people.

Sleep Diary
(to be filled in once you are fully awake in the morning)

Date	Mon	Tue	Wed	Thu	Fri	Sat	Sun
Before bed							
What time was your last food or drink?							
When was your last coffee or tea?							
How many alcoholic drinks did you have?							
During the night							
When did you go to bed?							
Did you find it hard to fall asleep?							
Did you wake in the night?							
If so, how often?							
And for how long?							

Date	Mon	Tue	Wed	Thu	Fri	Sat	Sun
What time did you get up?							
Estimate how long you slept							
Calculate your sleep efficiency							
Rate your sleep, 1 to 5							
How tired are you, 1 to 5?							
Thinking about yesterday							
Did you fall asleep unintentionally?							
How many cups of coffee did you have?							
Did you have an afternoon slump?							
Were you irritable?							
Did you exercise?							
Food and drink							
Are you eating more fiber-rich foods?							
Are you eating/drinking more fermented foods?							

How to estimate how long you have slept: If you have a sleep tracker it will do this for you. I have found that mine takes a lot of the stress out of recording my sleep patterns. It will also track your heart rate, making it more accurate than a device that simply records your movements. If you don't have a tracker, you simply take the time you went to bed, the time you woke, and take away roughly how long you were awake during the night. In other words, if you went to bed at 11 p.m., got up at 7 a.m., and were awake for two hours during the night, that means you were asleep for eight hours minus two hours = six hours.

How to calculate your sleep efficiency: In brief, take the amount of time you were asleep, turn it into minutes, and divide by the time you were in bed (again in minutes). In this case it would be 360/480 = 75%. Remember, you are aiming for 85%.

Purchase Any Sleep Aids

If you want to try some of the supplements I've written about, order them online now because they will take a while to arrive.

You can buy a light box from a shop or online, but make sure you choose one that provides 10,000 lux and has some decent ratings.

Measure Your Weight, Waist, and Blood Sugar Levels

I think it is very motivating to be able to measure the effect that improved sleep is having on your health, as well as your energy levels. I would expect your food cravings to decrease as you begin to sleep better, and for your waist and neck to shrink. So why not whip out a tape measure and measure both right now?

You measure your waist by wrapping the tape around

your belly button, not your hips. Why is waist size important? Because it is an indirect measure of your visceral, internal fat and one of the best predictors we have of future health. Ideally, it should be less than half your height (so if you are six feet tall your waist should be less than 36 inches).

Having too much fat around the neck is bad because it can affect breathing and lead to snoring and sleep apnea. If you are a man, it should measure less than 17 inches; for a woman, less than 16 inches. Losing an inch or more around your neck could make a huge difference to your sleep and the quality of your life.

I would also recommend you measure your blood sugar levels, as blood sugar control is badly affected by poor sleep. Sleep deprivation not only increases the stress hormone cortisol (which messes up your blood sugar levels) but also acts on hunger hormones like leptin and ghrelin, which will then cause you to overeat.

Around one in three adults in the US have prediabetes (raised blood sugar levels, not yet in the diabetic range) and most don't know they have it. Chronic insomnia raises your risk of prediabetes and type 2 diabetes, particularly if you are under 40.

The only way to tell if your blood sugars are abnormal is to measure them. You can ask your GP to test you, or you can do it yourself with a digital blood sugar monitoring kit, which you can buy quite cheaply at a pharmacy or online.

If they are currently raised, I would expect to see both your blood sugars and your blood pressure improve as you begin to sleep better.

If I'm overweight, should I prioritize improving sleep or losing weight?

If you are significantly overweight, you might be wondering whether to sort out your weight or your sleep first. You can do both at the same time. Losing weight will help you sleep better, while sleeping better will make losing inches around your waist and neck that much easier. Tackling both at the same time is certainly doable, though a challenge.

The good news is that the recipes in this book, based on a low-carb Mediterranean diet, discussed in Chapter 5, are not only great for helping your sleep but will also help keep you fuller for longer, which, in turn, will help you shed fat and keep it off.

Many people have told me that time restricted eating (TRE), see page 134, which I write about extensively in my most recent book, *The Fast800 Diet*, has helped them lose weight and improved their sleep. So do give it a go.

Get Cooking and Fermenting!

Feeding up your microbiome by eating more fiber-rich and fermented foods is an important part of this program, so take a look at our recipes and see which ones you fancy trying over the coming weeks. If your diet isn't currently fiber-rich, introduce these foods gradually as otherwise you will start producing a lot of gas!

You can buy fermented foods in the shops, but making your own sauerkraut is relatively easy and very rewarding. Just bear in mind that it will take time to mature. Details on page 262.

Sort Out Your Bedroom

Ensure that your bedroom is a place where you sleep and have sex, nothing more. So:

- If you have a TV in your bedroom, take it out.
- Replace any bright lights with bulbs that are softer and more diffuse.
- If you like the idea of music or white noise to fall asleep to, now is the time to get everything set up.

How's your mattress?

As a rule of thumb, you should replace your mattress every seven to ten years, but the life expectancy of a mattress varies considerably, depending on how good it was in the first place and how much of a pounding it has had. The main thing to watch out for is sagging. Take the sheets off and have a good look to see whether there is an obvious dip. If there is, a mattress topper will provide extra cushion and support and will be much cheaper than a new mattress.

What sort of pillow is best?

In theory, you should replace your pillow every couple of years. To test if it is time for a new one, try folding it in half and see if it springs open. If it doesn't, it probably won't be providing a lot of support for your head and neck.

If you do decide to replace it, what sort of pillow should you buy? According to the National Sleep Foundation (NSF), that depends on how you sleep.

People who sleep on their back will benefit from thinner pillows, "which help to limit stress on the neck." Stomach sleepers need a really thin pillow, or no pillow at all, to keep the spine straight and minimize stress on the lower back. For side sleepers (the most popular position), a standard pillow will do, though they might consider "placing a pillow between their knees or thighs to help maintain spinal alignment as they sleep."

Keep it dark

Finally, do make sure that your room is cool, dark, and quiet. If you have a clock, put it away, out of sight. Your cell phone should, ideally, be switched off or placed facedown on a table that is out of reach of your bed. You might want to invest in decent curtains or blackout blinds, particularly if you are a shift worker, though a sleep mask will be a good deal cheaper.

With all that sorted, it's time to take the plunge and get started. As I have said, this four-week program will be particularly helpful if you suffer from insomnia, but it will also benefit those of you who experience occasional disrupted sleep.

If you don't fancy sleep restriction, or you give it a go and find it too hard, do try the other key aspects of the program—practicing good sleep hygiene, changing the way you eat to create a sleepy biome, and doing your best to combat stress and anxiety to remain worry-free during the night.

Week One

SRT—How to Do It

First of all, you need to plan how many hours you are going to spend in bed for the next week—i.e., by how much you are going to restrict your sleep window.

Let's assume you currently go to bed at 11 p.m. and get out of bed at 7 a.m. But, though you lie in bed for eight hours, your tracker shows that you only sleep, on average, for six hours, giving you a sleep efficiency of 75%, which, you'll recall, is low.

Having established that you sleep for six hours, for the next week you are going to spend just six hours in bed each night. You will get up at the same time every morning, i.e., at 7 a.m. But instead of going to bed at 11 p.m., you are going to be going to bed at 1 a.m. (Similarly, if you have found that you are only asleep for 5.5 hours, then for the next week you are going to spend just 5.5 hours in bed; and so you will go to bed at 1:30 a.m.)

The four main rules, if you decide to try SRT, are:

1. Do not cut your time in bed to below five hours.
2. Stick to it rigorously.

3. Do not lie down, nap, or snooze during the day, and get your family to wake you if you do.

4. Do not drive or use machinery if you experience serious daytime sleepiness.

How long do I do it for?

SRT may radically improve your sleep in a couple of weeks, or it may take up to eight weeks to be fully effective. As your sleep efficiency improves, you increase the amount of time you spend in bed until you feel you are getting sufficient sleep.

If, as in the example on the previous page, you have cut the time you spend in bed to six hours, then you should find, after a few days, that although you are spending less time in bed, you are spending more of that time sleeping. So you are now in bed for six hours, but perhaps sleeping for five hours. In which case your sleep efficiency is now $\frac{5}{6}$ = 83%. Once it has reached 85% or better, for several nights in a row, you can move your bedtime to 20 minutes earlier.

Why 20 minutes? Why not 30 minutes? Or 40 minutes? There are different views, with some experts saying the increments should be in blocks of 15 minutes, and others suggesting blocks of 30 minutes work best; 20 minutes is a compromise.

Getting to 85% will probably take a week. If it doesn't start to improve after a week, you may need to cut your time in bed even further. In this example, you would go down to 5 hours and 40 minutes, going to bed at 1:20 a.m. If that doesn't do it, you could reduce it again, but I would be inclined to consult a doctor because it could mean you have a more complex problem. Remember, never go below five hours in bed.

Assuming, like most people, that you have reached your 85% target after the first week of the course, then for the second week you will be going to bed for 6 hours and 20 minutes. Your sleep efficiency may initially drop below 85% for a while. If it does, keep going with this new regimen until it is once more reliably back above 85%. This may take another week. When it is above 85%, add another 20 minutes to your sleep schedule. And so on, until you feel that you are getting not only enough sleep, but enough good-quality sleep.

You will know this is happening because you will wake up feeling more refreshed, you will have less daytime sleepiness, and you will be able to pass the Spoon Test (see page 59).

Most people will find four weeks of SRT is enough, but it can take up to eight weeks if you have serious long-term insomnia.

SRT is tough, particularly for the first week. You will probably feel more tired during the day than you do at the moment. You will feel moody and you may struggle to be sociable. You will almost certainly get the munchies.

You must tell your friends, family, and colleagues at work what you are doing, and why you are doing it, so they know why you are more sleepy, forgetful, and bad-tempered than normal. Try not to compensate by increasing your caffeine intake. And don't nap! This would be a bad time to go to the cinema, the theater, or anywhere warm and dark where you might nod off.

You may find it easier to adhere to the course with

professional support from a practitioner who is trained in delivering CBTi (cognitive behavioral therapy for insomnia). Find out more at https://www.babcp.com.

What should I do between 11 p.m. and 1 a.m.?

You might think that with all that extra time you have during the night, you will be able to do lots of useful or creative activities. But when I tried SRT I didn't do anything particularly productive. I read a lot of books and I watched a lot of TV. Although TV in the bedroom isn't a great idea, it is fine to watch it in the living room. But don't fall asleep, and do make sure you go to bed when you are supposed to.

What is the evidence that it works?

Sleep restriction is not new. It was first tested as a way of treating insomnia in the 1980s by an American psychologist named Arthur Spielman. In a now classic study, published in 1987,[67] he got 35 middle-aged patients who had been chronic insomniacs for more than 15 years and asked them to sleep restrict for up to eight weeks.

Before they started, the insomniacs were spending an average of 8 hours in bed but, despite taking sleeping pills, they were only sleeping for 5 hours and 20 minutes. In other words, they were tossing and turning for an average of 2 hours and 40 minutes every night. Their sleep efficiency was a miserable 67%.

Spielman didn't ask them to do anything else apart from cutting down to an average of 5 hours and 40 minutes in bed each night, slowly increasing that amount as the weeks went by.

The results were amazing (remember these were chronic insomniacs, most of whom had been taking sleeping pills for over 15 years).

Within a week they reported improved sleep. Over the course of the study their sleep efficiency improved from 67% to 87%. By the end of eight weeks they were spending 90 minutes less time in bed each night, but sleeping for longer than before. What I find particularly impressive is that the amount of time they spent awake, in bed, fretting, was down by almost 2 hours!

Even better, unlike drugs, doing sleep restriction had long-lasting results with no side effects. In a follow-up study carried out nine months later almost all of them had managed to keep their insomnia at bay.

Spielman's experiment has been repeated many times and a recent meta-analysis[68] showed very clearly that this technique works: cutting down time spent in bed really does reset the brain. People sleep more deeply, wake up less often, and feel much better during the day.

What are the benefits?

Firstly, your sleep improves really fast. People who've followed my advice and tried SRT say they're astonished how quickly they began to sleep much more deeply, which, in turn, improved their mood. They no longer worried about "not going to sleep" when they went to bed. Instead, they yearned for bed, fell asleep rapidly, and were much less likely to wake up during the night.

Interestingly, restricting sleep has also been shown to be a swift and effective way of treating depression. A recent review of 66 studies[69] looking into the impact of sleep restriction on different forms of depression found that around half of patients with depression (particularly those who were bipolar) responded well to it, though many found it hard to maintain. Researchers are currently looking at how to combine it with bright light therapy.

Sleep Hygiene

A quick reminder of the dos and don'ts.

Do:

1. Try some of our recipes—eating more fiber and fermented foods will help build a sleep-friendly biome, which can significantly boost your chance of a good night's rest.
2. Try TRE. Start with doing 12:12. That means not eating for 12 hours, for example between 8 p.m. and 8 a.m. Try to finish your evening meal at least three hours before you go to bed and avoid snacking before going to sleep.
3. Try cutting out all alcohol and cutting down on caffeine for a week and see if that helps.
4. Remember to fill in your sleep diary.
5. Get out of bed if you can't go to sleep and only return when you are tired.
6. Practice the breathing exercises, during the day as well as the night.
7. Expose yourself to bright light, whether outdoors or via a light box, for at least 20 minutes first thing in the morning.

Don't:

1. Have a TV in your bedroom.
2. Leave your phone beside your bed where you will be tempted to look at it.
3. Eat in bed! I met a woman who kept the drawer beside her bed stuffed with chocolate and wondered why she was sleeping so badly.

Week Two

SRT

If you are following the sleep restriction regime, this is the time to reassess. The week that has just passed may have been quite challenging, but hopefully you have succeeded in sticking to the course. If you have managed to avoid naps, stay awake till your later bedtime, and raise your sleep efficiency over 85%, then congratulations! Give yourself a hug.

Even better, you can now reward yourself with another 20 minutes in bed. Enjoy those extra 20 minutes and remind yourself why you are doing this: it is to reprogram your body and your brain and get yourself back on track. It is hard, but it will be worthwhile.

If your sleep efficiency hasn't yet improved, you can either stick to the current regimen or, as outlined above, reduce your sleep window by another 20 minutes, and see how that goes. For older people, hitting a target of 80% may be more realistic.

SRT-Lite

If you are falling asleep a lot in the day, or if it is really affecting your mood, full-on SRT may not be for you, in which case you could try something more mellow: simply try reducing your sleep window by just one hour. So, for example, if your normal sleep window is 11 p.m. to 7 a.m., try going to bed at 12 p.m.; and then gradually increase your sleep window as your sleep efficiency improves. It's a bit like losing weight. You can either go for rapid weight loss or attempt something more gradual.

Caffeine and Alcohol

Did you try cutting back on these two sleep disrupters, and did it help? Depending on how much you were drinking, I would expect an alcohol-free week to have made a difference. If you can manage another week, I would stick to it. After all, this is a four-week program and most people can manage a month without booze.

As for caffeine, hopefully you have found that you don't actually need it as much as you thought you did. Lots of people tell me that giving up one cup of coffee a day is no big deal and that, in the afternoon, a cup of tea, which is lighter on caffeine, fulfills much of the emotional need to give yourself a break.

TRE

Did you give this a go, and did it help? Getting used to TRE can take time, but most people find that they soon adapt. If you have found it relatively easy, you may want to extend your overnight fast from 12 to 14 hours, i.e., 14:10. If, however, you have found it a struggle or it has made your sleep worse, it is perfectly OK to put this aside for now and try again in a few weeks' time. It is also OK to break the rules on a couple of occasions a week, but try to maintain TRE five days a week.

Exercise

One thing I haven't written much about yet is exercise. By week two, you should be getting more sleep, which will put you in the mood for more exercise, or at least becoming a bit more active. If you are already exercising like crazy, this won't apply to you, but most of us really aren't doing as much as we should, either in terms of aerobic exercise (running, cycling, walking) or resistance exercises (push-ups, squats).

Becoming more active provides a huge range of health benefits, including improvements in mood and reductions in the risk of heart disease, cancer, and stroke. I don't like doing exercise for its own sake, and I hate the gym, so I have found ways to build it into my life. See the Appendix on page 267 for some simple resistance, aerobic, and HIIT (high-intensity interval training) exercises to help you get going.

Does it matter when I exercise?

It is best to exercise first thing, ideally in the early morning light and ideally in the fasted state (i.e., before breakfast). We're told that we shouldn't exercise in the evening because it interferes with sleep, but there seems to be very little evidence for that. The important thing is you do it, so don't worry too much about whether the time is right.

Will it help me fall asleep, straight off?

Probably not. Doing more exercise and being more active will undoubtedly help you sleep better. But not immediately.

In a small study,[70] researchers from Chicago asked 11 sedentary middle-aged women with insomnia to follow an exercise regimen for 16 weeks.

To start with, they met a sleep expert, who gave them advice on sleep hygiene (going to bed and getting up at the same time every day, etc.).

They were also introduced to a personal trainer, who spent the next four months guiding them through an exercise regimen. Four times a week, they had to do a 40-minute session in the gym, on a treadmill or an exercise bike, pushing themselves hard enough to get their heart rates up to about 120 beats per minute.

So what happened?

At the beginning of the experiment, the women were in bed for an average of 7 hours and 30 minutes, but only getting 5 hours and 54 minutes of actual sleep. By the end of the four months, they were getting 46 minutes more sleep per night.

The researchers found no direct link between the days on which the women had exercised and how they had slept that particular night. Instead, it was the other way around: after a bad night's sleep, the women found it much harder to get motivated to exercise and they got exhausted much faster. Sticking to their exercise regimen when they were tired became a struggle.

The moral is that you have to treat exercise as a way of life. The relationship between sleep and exercise is more of a virtuous circle than a quick fix.

The other issue to be aware of is that when you are tired, overweight, and sleeping badly, you will not want to exercise.

You will find every excuse to avoid it. The lure of the TV and the sofa will be irresistible.

You have to find ways to do it, even if you are not in the mood. Whatever you do, don't rely on willpower! Willpower is grossly overrated. Find something you actually like doing (Zumba? swimming? five-on-five soccer?) or make it something that is hard to avoid—park your car farther away from your house, for example, so you are less tempted to jump into it for short journeys, or make a regular arrangement to run or train with a friend.

Week Three

SRT

I hope you are still keeping your sleep diary and that, thanks to SRT, your sleep efficiency is up to roughly 85%. If it is, reward yourself with another 20 minutes in bed. If it isn't getting better, then keep going at the current level of sleep restriction for another week. And, remember, an SRT course can take up to eight weeks.

Techniques to Slow Down Your Racing Mind

Anyone who has wrestled with their sleep will be familiar with the feeling that your brain has gone into overdrive just as you want it to switch off. You can, of course, start using these techniques at any point in the program, but some people find that attempting too much, all at once, can be overwhelming. It's up to you.

An overactive brain is often the product of physical and psychological forces. Changing what you eat and when you eat should improve your microbiome and help reduce feelings of anxiety. But there are lots of other things you can do that will help.

As I mentioned earlier, one of the main things that keeps people awake at night is worrying about staying awake and the terrible consequences of not getting to sleep. Thoughts like: "I won't get to sleep and if I don't, then I will feel really tired at work tomorrow and get the sack."

It's important to realize that these thoughts are not real. They are not statements of fact. They can be challenged. You might want to give your negative thoughts a name, like "Donald." So when you have them, you can say: "That is just Donald sounding off again."

Challenging yourself sounds crazy, but it works.

The other key point to remember is that the negative thoughts you have at night are even less rooted in reality than the negative thoughts you have during the day. The filters are down, and you are more vulnerable to your inner demons.

You should also be aware that being sleep deprived will make you more prone to repetitive automatic negative thoughts (ANTs), along the lines of "I'm a failure, no one loves me," etc. Once your sleep improves, you will find that whenever you have ANTs they will be easier to stamp out.

Another way of approaching your catastrophic or negative thoughts is by imagining what a sympathetic friend would say to you if you were to share them. What would they say? How would they help to ground you?

As well as learning how to challenge these thoughts (which may well involve seeing a therapist), you can break the cycle by getting up and distracting yourself (reading a book, listening to music), or you can learn how to acknowledge them but then let them go by practicing mindfulness (see page 159).

Another technique that some people find helpful is something called "paradoxical intention," where you deliberately try to stay awake when you are desperate to go to sleep. So, rather than stressing yourself by thinking, "I must go to sleep now," you say to yourself, "I am enjoying being awake. I really am. Let's see how long I can stay awake for." It takes the pressure off, and by doing so it can, paradoxically, lead to sleep.

Breathing and progressive muscle relaxation

Although I like the idea of paradoxical intention, I personally find doing breathing and progressive muscle relaxation exercises more effective when I am struggling to sleep. As well as the 4-2-4 technique I described on page 100, you could try alternate-nostril breathing. This is the basis of a well-known yogic exercise, known as *nadi shodhana pranayama*, meaning "subtle energy-clearing breathing technique."

Start by breathing out through your mouth and then use your right thumb to close your right nostril. Breathe in deeply through your left nostril to a count of four. Really fill your belly. Now switch sides. Block your left nostril with your left thumb and breathe out fully to a count of four. Repeat ten times.

If you feel at all dizzy, which I did the first time I attempted this, you are trying too hard. Don't push yourself. This is supposed to be relaxing.

Progressive muscle relaxation is another exercise that anyone can do, but it would be a good idea to practice doing it a few times during the day to get the hang of it.

The idea is simple. While you are inhaling, contract one muscle group (for example, make a fist with your right hand) for five seconds, then exhale and at the same time release the tension in that muscle. As you do so, imagine those stressful feelings flowing out of your body. Then you give yourself a brief break (10 to 20 seconds), squeeze your eyes shut, and relax, before progressing through these muscle groups: right hand and forearm, right upper arm, left hand and forearm, left upper arm, belly, right thigh, and so on. There are videos at thefast800.com showing you how to do these exercises.

A word of warning, however: You shouldn't think of progressive relaxation as a way of solving your sleep problem. Once you start thinking of it as a solution, like taking a pill, you will start to ask yourself: "Is it working yet? Do I feel more relaxed? Will I sleep better?" That's where madness and insomnia lie.

Mindfulness

My go-to guy for mindfulness is Tim Stead, the author of *See, Love, Be* and a teacher at the world-famous Oxford Mindfulness Centre. According to Tim, one reason mindfulness can help with sleep problems is that it places a great emphasis on being in the moment; this is important for those of us who spend too much time lying there worrying about what has happened, or fretting about what is going to happen.

To get an impression of what mindfulness entails, try doing this exercise right now: sit up straight, close your eyes, and bring your attention to your breath, focusing on your chest rising and your lungs filling as your breath moves in and out of your body. No need to slow it down or speed it up; just see if it is possible to keep your attention on your own breath. If you notice that your mind has wandered, which it will, return to focus on your breathing. Don't dwell on the thoughts, simply notice them and let them drift away, like leaves on a stream.

The art of mindfulness is to keep doing this, but for progressively longer periods of time. If you can manage ten minutes once a day you will be doing well. Twenty minutes would be better.

Tim says that mindfulness helps because it encourages you to accept that you are awake, and that is fine. Once you let go, once you stop worrying about not going to sleep, then sleep will come. Like exercise, mindfulness is not a quick fix and it is unlikely that you will be able to do it successfully on your own. There are apps like Headspace, or Calm, that will guide you, or you could enroll in a mindfulness course.

Week Four

SRT

Once again, I hope you are filling in your sleep diary and that if you are doing SRT, your sleep efficiency is back up to near 85%, despite the fact that you are now spending more time in bed. If it is, reward yourself with another 20 minutes in bed.

By now you should definitely be seeing improvements in the quality of your sleep, and you will be finding it easier to fall asleep and stay asleep. You should be feeling less tired during the day, which, in turn, will motivate you to do more of the exercises that I have been recommending.

As I have said, most people will find that four weeks of SRT is enough to mend their sleep problems, although you can continue for up to eight—it very much depends on how you are getting on.

Looking after Your Old Friends

If you have been eating meals from the recipe section in this book, I would also expect your gut microbiome to have changed radically, and for the better. Your levels of "good" bacteria should have increased, reducing inflammation and making you feel more cheerful, while the "bad" ones, which cause inflammation, will have been displaced. So keep munching those legumes! Remember that quite apart from the positive impact that these foods have on your sleep, they will also help cut your risk of type 2 diabetes, heart disease, and dementia. Treat this way of eating as a way of life, not just a quick fix when it comes to improving your sleep.

Eating for Better Health and Weight Loss

As we have seen, if you are overweight, even losing a bit of fat will make a huge difference to your sleep and the quality of your life.

At the beginning of the program, I recommended you write down your weight, waist size, neck size, and blood sugar levels. Have they improved?

I wrote that ideally your waist should be less than half your height, and if you are not there yet, are you getting a bit closer? If your neck size has decreased, has that had an effect on your snoring? How about your blood sugar levels? Were they a problem before you started, and if so, have they fallen? If you have blood sugars that put you in the diabetes range, I would strongly recommend you check out thefast800.com website for more information on the benefits of rapid weight loss.

Becoming More Active

Have you been doing more exercise? I would hope that you have started both the resistance exercises and doing HIIT a few times a week. As your sleep improves, you will find doing exercise much more enjoyable. It really is key to healthy living. Do keep going.

Keep Calm and Carry On

It is early days, but if you have been practicing the breathing and muscle relaxation exercises, you should soon start reaping the benefits. Mindfulness is demanding, at least initially, and the benefits take longer to become apparent. But if you are stressed and haven't yet decided to give it a go, then I would strongly recommend trying an app or signing up for a course.

• •

The Fast Asleep Program in Brief

Countdown to Sleep

- Clear out your bedroom and order any equipment or supplements (sleep trackers, light box, melatonin, Bimuno).
- Establish good bedtime habits.
- Start keeping a sleep diary.
- Eat more Mediterranean-style foods.
- If you're going for the sleep reboot, use your sleep diary or tracker to work out your sleep efficiency.

Week One

- To reboot your sleep, reduce the amount of time you spend in bed each night so it matches the time you actually spend sleeping each night.
- If you don't think you can handle a full reboot, simply follow the other suggestions in this book and try reducing your sleep window by just one hour.

Week Two

- If your sleep efficiency has improved to 85%, add 20 minutes to your sleep window.
- If you're not seeing any improvement, keep going for another week.
- Incorporate TRE (14:10) into your life.
- Add in some exercise or at least increase your activity levels.

Week Three

- If your sleep efficiency really doesn't improve after two weeks of sleep restriction, cut the time you spend in bed by another 20 minutes. In other words, if you started by spending six

hours in bed, you will now have to go down to 5 hours and 40 minutes. (NB: Never go below five hours in bed.)
- Practice challenging your catastrophic thoughts.
- Do breathing and muscle relaxation exercises.

Week Four
- Your sleep efficiency should by now have improved significantly, and you should be feeling much better during the day. If you feel that you are getting sufficient sleep, then you can ease up on the sleep restriction. If not, keep going.
- Remember to get outside first thing for at least half an hour to soak up the light.
- Keep eating that Mediterranean-style food.
- Continue with the mindfulness, breathing exercises, and other forms of stress-busting and relaxation.

. . . .

Once you have got your insomnia under control, the best way to ensure you go on sleeping well is to follow these eight simple rules:

1. Stick to a regular sleep window, i.e., a regular wake-up-go-to-bed routine.
2. Use the sleep tips from Chapter 4 that work for you.
3. Manage your stress by practicing mindfulness and breathing exercises during the day.
4. Get out of bed if you can't go to sleep and don't get back in until you feel tired.
5. Expose yourself to bright light (daylight or a light box) first thing in the morning.

6. Remain active and do plenty of resistance exercises like push-ups and squats.
7. Eat a Med-style diet with some fermented foods.
8. Try to keep your belly fat down by aiming for a waist that is less than half your height.

7.

HOW TO MANAGE SHIFT WORK AND JET LAG

Most of this book has been aimed at people who follow the normal pattern of waking and sleeping—people who go to bed at night and get up in the morning. But there is a large and growing section of the population who don't do this. They work at night and sleep, if they can, during the day. They are shift workers and they face a very particular set of challenges.

We are not designed for shift work. Our remote ancestors rose at dawn, went hunting or grubbing for food, spent the day largely outside, then retired to their caves for sex and sleep. Their lives and their all-important internal clocks were almost entirely driven by the movements of the sun.

Then, in the 20th century, we saw the invention of the electric light bulb and the jet engine, both of which had a major impact on our circadian clocks.

The jet engine meant that we were now able to get to the other side of the world in 24 hours, which is way too fast for our ancient body clocks to adjust to; while the creation of intense artificial light meant that not only were teenagers able to stay up until the small hours, but an increasing number (roughly 20%) of the population became shift workers, toiling away when their body clocks were screaming at them to sleep.

Jet lag and shift work have a lot in common. In both cases your internal body clock is thrown out of sync with the external world, and this has a number of unfortunate consequences.

In the case of jet lag, the impact can be unpleasant, but it tends not to be long-term (unless you do an awful lot of traveling). For shift workers it can be truly life changing. Fortunately, there are things you can do to make shift work more tolerable, and I will come to those in a moment. First, I want to look at jet lag.

How to Overcome Jet Lag

Lots of flying is bad for the planet and it is not good for the brain. A study found that putting female hamsters on a brief jet lag regimen damaged the hippocampus, an area of the brain that is important for learning and memory.[71] The effects on memory persisted for weeks after the jet-lagged hamsters had returned to their normal sleeping pattern.

Jet lag can make you do stupid things. Former US president George W. Bush memorably tried to blame jet lag for an embarrassing press conference in Beijing, which he tried to cut short by attempting to exit through what was plainly a locked door. After tugging repeatedly at the door, he admitted to the gathered press: "I was trying to escape. Obviously, it didn't work."

I haven't done anything quite that publicly embarrassing, though I did once manage to walk out of my hotel room in the night, semi-naked, thinking I was heading for the bathroom. By the time I realized what I had done, the door had slammed shut behind me. I had to sneak down to reception and get another key.

Since jet lag is caused by the imbalance between your internal body clock and your new time zone, the journeys that produce the most brutal effect are, not surprisingly, the ones that involve flying through multiple time zones. In my case, that means flights to places like Australia and the US, which I have to do occasionally for business.

Before I found ways of coping, jet lag would leave me feeling dazed, hungry, and irritable, madly craving carbs at crazy times of the day and night.

The good news is that there are things you can do to reduce the impact of jet lag. But they require a bit of forward planning.

Sleeping Pills and Melatonin

If you don't already have them, see if you can get prescriptions for sleeping pills and melatonin from your GP. Although doctors are understandably reluctant to prescribe sleeping pills for long-term use, they are more sympathetic if you just need some to get you through jet lag. I use zopiclone. (Zopiclone is not available in the US, but zolpidem is a good alternative.)

As I mentioned on page 93, melatonin is an important hormone for resetting your internal clock. A Cochrane review,[72] which looked at nine studies involving nearly 1,000 people, concluded: "Melatonin is remarkably effective in preventing or reducing jet lag, and occasional short-term use appears to be safe. It should be recommended to adult travellers flying across five or more time zones, particularly in an easterly direction, and especially if they have experienced jet lag on previous journeys. Travellers crossing 2–4 time zones can also use it if need be."

I find this pretty convincing, as Cochrane reviews are widely regarded as the best when it comes to assessing medical evidence.

The Argonne Diet

Widely used by the US military, this is a way of combating jet lag created in the 1980s by Dr. Charles Ehret, a researcher at the Argonne National Laboratory, near Chicago. Dr. Ehret, an expert in circadian rhythms, discovered that you could reset your internal clock faster if, for three days before flying, you alternated feasting and fasting. His regimen was simple:

- On day one, a fast day, you have to restrict yourself to 800 calories. You can find suitable recipes in my book *The Fast800 Diet*, or go to thefast800.com.
- On day two, a feast day, you eat a high-protein breakfast, an above-average-size lunch, and have an early dinner. Don't drink coffee after 5 p.m.
- On day three, the day before your flight, you have another 800-calorie day.
- On the day of your flight, you don't break your fast till your destination breakfast time. In other words, if you are in London, traveling to New York, your first meal of the day would be at 1 p.m. UK time (8 a.m. in New York). If you are traveling to Sydney, on the other hand, you would not eat until 7 p.m.

So does it work? In 2002, the US military decided to put the Argonne diet to the test.[73] They split 186 soldiers, traveling to South Korea for military duties, into two groups; half did the Argonne diet, the others ate normally. They found those who followed the Argonne diet were seven and a half times less likely to experience serious jet lag than those who ate normally.

The Fast Version of the Argonne Diet

If the Argonne diet sounds a bit tough, there is a simplified version you can try, developed by researchers at the Beth Israel Deaconess Medical Center in Boston. It is the one I most commonly use.

The idea is simple: you do a short fast on the day that you fly. It is based on the fact that, like light and darkness, when you eat has a powerful effect on resynchronizing your body clocks.

When you are "fasting" you should drink plenty of water and herbal tea, but no alcohol or caffeine. Below is an example of what I do to minimize jet lag, based on extensive personal experience and lots of chats with air crews and sleep experts.

1. On the day of the flight

Firstly, I always pack a sleep mask, wax earplugs, an empty water bottle, noise-canceling headphones, an inflatable travel pillow, and a good book.

If I am heading to New York—five hours behind the UK—I like to get an early afternoon flight. When I get up, I skip breakfast and aim not to eat anything until at least 1 p.m. (8 a.m. New York time).

At the airport, I set my watch to New York time, then walk around a lot. I do squats and push-ups and use airport chairs to do triceps dips. I also drink a lot of water. I don't drink any alcohol or caffeine in the airport or on the plane.

If I get a 1 p.m. flight, I will normally have a light lunch on the plane, and then nothing more until I have a high-protein evening meal when I get to my hotel in New York (which is normally around 6 p.m.). I go to bed at 9 p.m., having swallowed a melatonin capsule and a sleeping pill.

2. The next morning

I get up at around 6 a.m. and when it is light I go for a 30-minute walk (or run, if I'm feeling particularly energetic). This is partly to get some exercise but also to grab some morning light. Then I have a protein-rich breakfast (omelets, scrambled eggs, that sort of thing) and get on with my day.

I normally take melatonin on the second night, and sometimes a sleeping pill. I find that within a day or two I can sleep happily without them.

3. Returning home

On the way back to London from the US, I get an early evening flight, after a good lunch at 2 p.m. I skip dinner on the plane and, with the aid of melatonin and a sleeping pill, try to get to sleep as quickly as possible. I also skip breakfast on the plane.

When I get back home, after a 14-hour fast, I have breakfast. I go for a brisk midmorning walk or run, to perk myself up. I have a light lunch, a protein-rich evening meal, and go to bed at my normal time, after taking a melatonin capsule and a sleeping pill. This usually works brilliantly.

Q&A

Why is it harder flying east than west?

It is mainly because it is easier to stay up and go to sleep later than normal, rather than go to sleep much earlier than normal. If you have flown west, from London to New York, and it is 9 p.m. in New York, you should find it easy to go to sleep because your internal clock is telling you that it is 2 a.m.

Going east, from New York to the UK, means that when it is 11 p.m in London your body is telling you it is really only 6 p.m., so why on earth are you going to bed? That's when I really need a sleeping pill.

What about naps?

Personally, I don't do naps, because I need the buildup of sleep pressure to put me to sleep and keep me asleep. If you find them refreshing then that is fine, but no more than 40 minutes.

How to Manage Shift Work

While jet lag is unpleasant, shift work can be deadly. There is a long list of terrible things that prolonged shift work can do to the human body, ranging from increased risk of heart disease, type 2 diabetes, cancer, and obesity, through early menopause, depression, and divorce.

One shocking statistic that I came across is that in the US, more firefighters are killed in traffic accidents and from heart attacks than in fires. Although firefighters (like most shift workers) are likely to have a major sleep disorder, few seem to realize it.

A study of 7,000 American firefighters[74] found that 37% had at least one major disorder, the most common being obstructive sleep apnea. The vast majority of them (80%) were undiagnosed and didn't realize that this put them at twice the risk of heart disease or diabetes. They were also more than three times as likely to suffer from depression and anxiety.

A similar study, which looked at American police officers, showed similar results, with about 40% having significant sleep problems.[75]

What is their life like? In the course of researching this book, I had a fascinating chat with Joe, who is 48 years old and has been a firefighter for 18 years. Like most firefighters in the UK, he works two day shifts, followed by two night shifts, then has four days off. On his day shifts, he works from 9 a.m. until 6 p.m.; then he does two night shifts, when he works from 6 p.m. until 9 a.m. When he's doing the night shifts, he gets the chance to have a bit of shut-eye, but this is unpredictable, and it is always noisy at the fire station.

Joe is a bit of a foodie, so when he is doing a night shift he tries to eat healthily. He brings his own meals to work, featuring high-protein foods such as poached eggs, smoked salmon, and avocados. There is a canteen in his fire station, which does provide food, but it is mainly pasta and other high-carb, rather stodgy foods.

Unlike most of the others, who often snack, Joe tries not to eat again until lunchtime the next day. He has been a fan of time restricted eating (TRE) for many years, and he finds that eating during his night shift makes him feel sluggish.

If things are quiet, the team is allowed to sleep in a dormitory, but they are expected to be up and out of the building within 90 seconds of the alarm going off. So Joe sleeps lightly, half awake, anticipating the ringing of the alarm: "I always wake to the soft click, which happens just before the alarm itself goes off."

After he has returned home from a night shift, Joe prefers not to go to sleep, because this leaves him feeling tired and grumpy. His main way of coping is by doing lots of exercise. He finds doing night shifts has gotten harder as the years have gone by.

What I found startling, when I talked to firefighters like Joe, and to police officers, nurses, and paramedics, is that none of them get any specific advice on how to cope with being a shift worker. They are just left to get on with it.

My son Jack is also a shift worker. He is a junior doctor, and every few weeks he has to do a couple of nights on call, from 9 p.m. to 9 a.m. Jack has a nap in the afternoon before work, and a meal in the early evening. He also takes food into the hospital to eat before midnight, as the only alternative is something out of a vending machine. And, like Joe, he practices TRE, trying not to eat between midnight and 9 a.m., when he finishes work. Instead, he drinks lots of water and tea.

He also takes a sleeping bag into the hospital with him and, depending on how busy he is, sometimes manages to get a bit of sleep on the floor of the doctors' office. He finds that a short nap in the early hours means he feels less tired when driving home after a night on call.

What Can You Do to Counter the Effects of Shift Work?

In recent years, there has been a lot of research into shift work and some things are very clear.[76]

If you are an employer:

1. You should be aware that employees over the age of 45 years are less able to adapt to shift work, and the impact on their bodies and brains is greater. I had a chat with Dave, a nighttime security guard, who says he relies on Red Bull, junk food, and cigarettes to stay awake. He is in his mid-fifties and has put on 33lb (15kg) since he started doing night work two years ago. He has type 2 diabetes, high blood pressure, and sleep apnea, and gets about five hours' sleep a night. Not surprisingly, he often falls asleep at work.

2. If you are creating a shift-work schedule, then do take your employees' chronotype into account. Don't make larks do the late shift, or owls the early shift, if you can avoid it.

3. If your employees have to do a rotational shift, it should always go in a clockwise direction (start them off doing days, then evenings, then nights), because it is much easier to adjust to this than random shifts or shifts that go counterclockwise.

4. Do try and provide a place where employees can have a short nap, even if it is only 20 minutes. A study of engineers in New Zealand[77] showed that getting a 20-minute nap during the night shift significantly improved their performance, while a study of American nurses found that a 20-minute nap meant they were less drowsy on their drive home. This study also found that many hospital managers, who don't work nights, are very resistant to providing napping opportunities.

If you are a night shift worker:

1. Before your shift begins, take a long nap. If you can, try to get the majority of your sleep in the hours leading up to starting work. There is evidence that if you are working a night shift, say from 11 p.m. to 7 a.m., it is better to have an evening sleep (from 2 p.m. to 9 p.m.) rather than a morning sleep (8 a.m. to 3 p.m.). In one study,[78] shift workers who slept in the afternoon made fewer mistakes at work than the morning sleepers.

   If you are driving to work, you should try to wake up at least an hour before heading off, as it can take that long to become fully alert.

   Put together a bag with healthy food and drink to take with you. The kind of food available from vending machines is likely to be high in sugar, saturated fat, and salt, and low in fiber and nutrients. When you eat junk food, you pack in the calories and are soon hungry again. Plus, the fact that you are eating at night, when your body will find it hard to process, means the bad stuff will hang around in your system longer. So try to eat your main meal before midnight, and take nuts, apples, and pears to snack on if you can't endure a long night without anything.

Take a bottle of water from your fridge to sip during overnight shifts. Do not drink caffeinated fizzy drinks.

2. While you're at work, during the early hours of your night shift, make sure you are getting some bright light. If the place where you are working is dimly lit, bring in a light box to give yourself a 20-minute blast. It is more effective and certainly healthier than caffeine at night.

Do try TRE. There is a lot of evidence that eating when your body thinks it should be asleep is really bad for the heart. There are a couple of major studies going on at the moment in the US and Australia looking at the benefits of limiting the hours within which you eat while doing an overnight shift. Researchers from Monash University, for example, are asking people to avoid eating between 1 a.m. and 6 a.m., to see what impact that has on their risk of heart disease.[79]

If you get the chance, and there is somewhere you can safely do it, get in a 20- to 40-minute nap at some point during your shift.

3. When you are traveling home after a night shift, try to car-share, take public transport, walk, or cycle. If you have to drive, a short nap before heading off in the car may also help.

On the journey home, you could try wearing dark glasses. The idea is to avoid morning light, where possible.

Some people go straight to bed when they get home, and if you have a family with young children, this makes sense because they will probably be at school. However, research so far shows that sleeping in the afternoon seems to be better.

To get a decent sleep, you will probably need wax earplugs or white noise, a sleep mask, and a prominent "Do Not Disturb" sign on the bedroom door. You will also need to follow the sleep hygiene tips I recommended in Chapter 4: i.e., have a fixed time when you go to sleep and wake up; a strict countdown to your bedtime routine; no caffeine or alcohol before bedtime, etc.

Will Taking Melatonin Help?

The short answer is that it may. The Cochrane reviewers concluded that the use of melatonin by night shift workers increased the amount of sleep they got by 24 minutes.

Do All Shift Workers Have Problems?

No. There are some people who cope well with shift work, but being older and female seems to make it harder to adapt. Apart from the impact on your body, it can also mess with your personal life. If you or your partner are shift workers, you are six times more likely to get divorced than if you work days.[80]

One of the reasons for such high rates of divorce is that

shift workers find it hard to take part in "normal" family life and plan family events.

As Sue, a nurse who works nights, told me: "Working nights means you have time at home, in the day, so it's great for going to the bank or shopping. The trouble is that most school events, like plays or meeting teachers, happen in the evening. My husband, who is a paramedic, also works nights. We leave each other notes in the kitchen, but sometimes it feels like days go by without us really talking to each other."

People who don't adapt quickly develop a condition called shift work sleep disorder (SWSD). Common symptoms of SWSD include:

- Feeling excessively sleepy, both on and off the job
- Having difficulty concentrating
- Finding it hard to go to sleep and waking up feeling tired
- Feeling depressed or moody
- Having trouble with close relationships, such as with friends and family

The American Sleep Association recommends, among other things, that people who suffer from SWSD try light therapy, "which involves sitting near a light box for a prescribed amount of time," and melatonin, though they do emphasize that it is important to talk with your doctor before taking it. Your doctor may also prescribe you sleeping pills or modafinil.

Modafinil is a wake-promoting drug, widely used by the American military to keep their troops alert. Its main medical use is to treat narcolepsy, a rare brain condition that causes people to suddenly fall asleep at inappropriate times. People with narcolepsy who I've interviewed have told me extraordinary stories of falling asleep on roller coasters, in the middle of dinner, or even when out riding a horse.

Studies have shown that taking 200mg of modafinil an hour before a night shift can significantly improve your performance without affecting your sleep the following day.[81] As well as perking you up, modafinil has been shown to reduce long-term memory damage in shift workers with SWSD.

Modafinil is a prescription-only drug, and it has far more side effects than melatonin. Despite this, it is widely, and illegally, used as a cognitive enhancer or "smart drug." It is particularly popular with university students; surveys suggest that up to 20% use it around exam time. However, taking modafinil in the hope that it will improve your exam results is almost certainly counterproductive; a recent study conducted with healthy volunteers showed that students taking modafinil did worse in memory and cognitive tests than students taking placebo pills.[82]

I took modafinil a few years ago as part of a sleep deprivation experiment and it certainly kept me wide awake. However, as I've just pointed out, modafinil can have significant side effects, including the risk of developing an allergic reaction. Which is what happened to me. I developed such a severe allergic reaction that I had to go to the hospital, where they gave me a big dose of steroids to dampen down my immune system. In retrospect, it is one of the scariest self-experiments I have done. So, never again.

Summary

- Problems with jet lag and shift work are caused by the fact that your circadian clock is out of sync with the world outside. You are trying to sleep when your body clock is saying you should be awake, and vice versa.
- Jet lag can be improved by following either the Argonne diet or a modified version that requires a 14-hour fast.
- The rule is that you don't eat until the equivalent of breakfast time at your destination. So if you are flying from London to New York, you skip breakfast and don't eat until at least 1 p.m. (8 a.m. New York time).
- Doing this will help speed up how quickly your body is able to adjust to local time.
- Shift work is harder to treat, but there are things that can help, such as having your main sleep in the hours running up to work and trying to practice TRE while at work.

GOOD NIGHT FROM ME

Insomnia is a serious condition that affects hundreds of millions of people worldwide. In recent years, there's been mounting concern about the impact that shift work and sleep deprivation have on our brains and bodies. The sleep industry is huge and growing fast, yet few of the things they are selling actually work.

The methods I have outlined in this book, however, do have strong scientific support and I am sure they will help most people who have a sleeping problem. If you have suffered from insomnia for a long time, as I have, it will be harder to crack than if you occasionally have a bad night, but it can be done!

If, despite working on your sleep program as suggested, you are still struggling, it may be worth discussing it with your GP.

Long-term insomniacs are also at greater risk of being tipped back into their sleepless state by stressful life events. It won't necessarily happen, but if it does, don't worry. Just redo the Fast Asleep program and in a short while things should once more be under control. Good luck! And do visit our website at fast-asleep.com.

Black Grapes with Yogurt and Almonds (page 186)

Sweet Potato Savory Muffins with Feta (page 190)

Chicken Noodles with Bok Choy (page 198)

Curried Lentil Soup with Turmeric (page 200)

Dr. Tim's Hot-Smoked Salmon Salad (page 203)

Griddled Eggplant with Feta and Pine Nuts (page 204)

Shrimp and Red Cabbage Slaw on Sourdough (page 216)

Black Bean Salad with Lime and Avocado (page 219)

Chicken and Vegetable Tray Bake (page 232)

Meatballs with Feta and Eggplant (page 235E

Mushroom, Chickpea, and Kale Curry (page 237)

Smokey Fish Gratin (page 243)

Braised Red Cabbage with Walnuts and Apple (page 250)

Seeded Rainbow Salad (page 252)

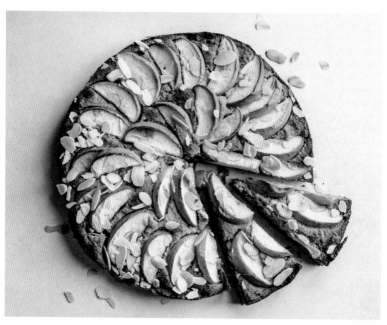

Dorset Apple Cake (page 256)

Nutty Berry Crunch (page 261)

RECIPES

by Dr. Clare Bailey and Justine Pattison

As a GP, I have been struck by how many of my patients report sleeping better when they follow my advice and switch to a moderately low-carb, high-fiber Mediterranean-style diet. This is partly because of the effect it has on their waistlines and their necks and therefore on how much they snore, but it is also because of something most of them are unaware of: the impact that a change in diet is having on their microbiome.

The recipes here have been carefully chosen to help you boost nutrients and increase your fiber intake, all for the benefit of looking after your microbiome so that it looks after you.

Dr. Clare Bailey

Note: All calorie counts are for one serving.

BREAKFAST AND BRUNCH

These easy-to-prepare breakfasts are filling and tasty, and will help you get going in the morning and keep your energy levels up.

Black Grapes with Yogurt and Almonds

With a boost of gut-friendly live yogurt and the anti-inflammatory resveratrol in the grapes, this is a great way to start the day.

255 calories: Serves 1

½ cup full-fat live Greek yogurt or plant-based yogurt
3 to 4 black or red seedless grapes, halved (about 2 ounces)
½ ounce flaked almonds, toasted
1 teaspoon honey (optional)

Spoon the yogurt into a bowl or glass tumbler. Top with the grapes and flaked almonds, drizzle with honey, if using, and serve.

COOK'S TIP: If you can't buy toasted almonds, simply toast your own. Scatter the almonds in a dry frying pan and cook for 2 to 3 minutes over medium heat, tossing regularly, until lightly browned.

If making this a portable breakfast, assemble the ingredients in a lidded container and keep chilled until you are ready to eat.

Oaty Nutty Shake

A delicious, creamy oatmeal shake will help lower your cholesterol, as well as giving you a good portion of healthful cashews.

295 calories: Serves 2

1½ ounces plain cashews (not roasted)
4 tablespoons steel cut oats (about 1 ounce)
1½ cups whole milk or plant-based milk
1 teaspoon pure maple syrup
¼ teaspoon ground cinnamon

1. Put the nuts in a bowl and cover with cold water. Refrigerate for 4 to 6 hours to soak and soften. (If you have a really powerful blender, you can skip the soaking stage.)
2. Drain the nuts and transfer them to a large bowl. Add the oats, milk, maple syrup, and cinnamon and blitz with an immersion blender until as smooth as possible, adding a little extra milk if needed to keep the shake liquid. (You could do this in a blender or food processor.)
3. Pour into two glass tumblers to serve.

Rosy Overnight Oats

Oh, the joy of stumbling into the kitchen and being able to dig straight into this creamy, oaty breakfast pot! It's also a great way to get a good dose of fiber.

330 calories: Serves 2

> 1 small apple, quartered, cored, and coarsely grated
> About 4 ounces frozen mixed berries or fresh berries, in season
> ¼ cup steel cut oats, preferably Irish oats, which tend to be larger than others
> 1 ounce flaked almonds, toasted
> 8 dried apricots, coarsely chopped
> ⅓ cup full-fat live Greek yogurt or plant-based yogurt
> ½ cup whole milk or plant-based milk

1. Place the apple in a large bowl and then stir in the berries, oats, almonds, apricots, yogurt, and enough milk to reach a soft, creamy consistency (the mixture will thicken as it stands). Cover and refrigerate for several hours or overnight.
2. To serve, divide between two small bowls. Add extra milk, if needed. This will keep well for up to 2 days if refrigerated. If taking to work or school, store in lidded containers and keep refrigerated.

COOK'S TIP: If you can't buy toasted almonds, see page 186 for a tip on how to toast them.

Scrambled Eggs with Kimchi

This is Michael's favorite breakfast at the moment—and a great boost to a healthy, sleep-enhancing microbiome.

345 calories: Serves 2

> 1 tablespoon unsalted butter or olive oil
> 4 large eggs, well beaten
> Flaked sea salt and freshly ground black pepper
> 2 slices seeded sourdough bread (about 1½ ounces per slice)
> 3 tablespoons kimchi (Korean pickled cabbage) or sauerkraut

1. Melt the butter in a medium nonstick saucepan.
2. Add the eggs, season lightly with salt and pepper, and cook over low heat for 2 to 3 minutes, stirring regularly, until lightly set.
3. Meanwhile, toast the bread on both sides and divide between two plates. Top with the scrambled eggs and serve with the kimchi alongside.

COOK'S TIP: Kimchi is available in most large supermarkets and in health food shops. You could try making it yourself (see our recipe in *The Fast800 Diet*, page 204).

Sweet Potato Savory Muffins with Feta

Enjoy these as breakfast or brunch with a large salad, either hot from the oven or the next day at room temperature, removing them from the refrigerator 30 minutes before serving or warming through in the microwave. Keeping the skin on the sweet potato not only saves time because there is no peeling, but adds lots of extra nutrients and fiber.

335 calories: Makes 6 muffins

5 tablespoons olive or rapeseed oil, plus extra for greasing

1 sweet potato (about 8 ounces), scrubbed well and cut into roughly ¼-inch chunks

Flaked sea salt and freshly ground black pepper

1 medium onion, coarsely chopped

1 tablespoon chopped fresh thyme leaves (or ½ teaspoon dried thyme)

½ cup ground almonds

⅓ cup wholemeal self-rising flour

½ teaspoon baking powder

1 large egg, lightly beaten

½ cup whole milk

4 ounces feta or goat cheese, cut into roughly ¼-inch cubes

About 1 ounce Parmesan cheese, finely grated

1. Preheat the oven to 475°F. Generously grease a 6-cup non-stick muffin pan with oil.
2. Scatter the sweet potato chunks in a small roasting pan. Toss with 1 tablespoon of the remaining oil, season with salt and pepper, and roast for 10 minutes.
3. Remove from the oven and, using a long-handled spoon,

turn over the pieces of sweet potato. Scatter the onion and thyme leaves over the potato and roast for 15 to 20 minutes longer, or until the sweet potato and onion are tender and lightly browned.

4. Meanwhile, mix the ground almonds with the flour and baking powder in a bowl and whisk in the remaining oil, the egg, and the milk.

5. Stir the roasted vegetables into the batter and divide among the cups in the muffin pan. Dot with the cubes of feta and sprinkle with the Parmesan. Bake for 20 to 25 minutes, or until puffed up and golden brown. Eat warm or at room temperature.

COOK'S TIP: If your muffin pan is losing its nonstick properties, line cups with circles of parchment paper.

LIGHT MEALS

Most of these easy meals are low in calories, as well as being quick and easy to prepare. Perfect for a light lunch, with lots of fiber for those microbes to chomp on.

Cauliflower and Roasted Red Pepper Soup

This creamy soup is full of soluble fiber and loads of flavor. For a more filling meal, serve the soup topped with crumbled blue cheese or scattered with toasted seeds.

120 calories: Serves 5 or 6

3 tablespoons extra-virgin olive oil

1 large onion, coarsely chopped

4 celery stalks, trimmed and thinly sliced

1 garlic clove, thinly sliced

1 medium cauliflower, trimmed and cut into roughly 1½-inch chunks, including the stem (about 1 pound prepared weight)

6 ounces jarred roasted red peppers, drained

6⅓ cups chicken or vegetable stock (fresh or made with 1½ stock cubes)

Flaked sea salt and freshly ground black pepper

Handful of flat-leaf parsley or fresh cilantro leaves, coarsely chopped, to serve (optional)

1. Heat the oil in a large deep saucepan and when hot, gently fry the onion and celery for about 5 minutes, or until softened and beginning to brown, stirring regularly. Add the garlic and cook for a few seconds more, stirring constantly to prevent burning.
2. Add the cauliflower, red peppers, and stock to the pan and bring to a boil. Reduce the heat, cover loosely with a lid, and simmer for about 20 minutes, or until the cauliflower is soft, stirring occasionally.
3. Remove the pan from the heat and blitz the soup with an immersion blender until smooth. (You could do this in a food processor.)
4. Season to taste with salt and pepper. Warm through gently, stirring regularly, adding a little extra water if needed. Ladle into warmed bowls to serve. Garnish each serving with parsley, if using.

Jerusalem Artichoke Soup

This soup is a perfect way to enjoy one of the ultimate sleep-enhancing, high-fiber vegetables. Keep the skin on to retain the best nutrients. If you aren't familiar with Jerusalem artichokes (which sometimes are called sunchokes), serve small portions to begin with, as the amount of fiber may create wind at first.

325 calories: Serves 4

4 tablespoons extra-virgin olive oil
2 medium onions, finely chopped
2 garlic cloves, crushed
1¾ pounds Jerusalem artichokes, scrubbed well and cut into
    roughly ¼-inch slices
3 cups chicken or vegetable stock (fresh or made with
    1 stock cube)
Flaked sea salt and freshly ground black pepper
4 tablespoons mixed seeds, such as pumpkin, sunflower,
    sesame, and flax, ideally toasted

1. Heat the oil in a large saucepan. Add the onions, garlic, and artichokes, cover with a lid, and cook gently for 15 minutes, or until the vegetables are very soft, stirring occasionally.
2. Add the stock and bring to a simmer. Cook, uncovered, for 5 minutes, stirring regularly.
3. Remove the pan from the heat and blitz with an immersion blender until smooth. (You could do this in a food processor.)
4. Season with salt and pepper to taste, and gently warm

through. Ladle the soup into warmed bowls and serve in small portions, sprinkled with the seeds.

COOK'S TIP: Toast the mixed seeds in a dry frying pan over low heat for 2 to 3 minutes, or until they darken a shade or two, stirring constantly. For a richer-tasting soup, add a little whole milk or cream when reheating.

This soup freezes beautifully so, if you don't think you'll get through it all in a couple of days, pop the rest into the freezer.

Anti-Inflammatory Chinese-Style Chicken Broth

This is a gut-soothing chicken broth, full of vital nutrients. You can enjoy it as a clear soup, or see page 198 for a fuller version with added fiber, vegetables, and protein.

170 calories: Serves 4 (Makes 6⅓ cups)

- 1 pound organic chicken wings and/or leftover roasted chicken
- 1 medium onion, quartered
- 1 medium carrot, scrubbed well, trimmed, and sliced
- 2 celery stalks, trimmed and cut into lengths roughly ¾-inch long
- 4 garlic cloves, halved
- 2 ounces fresh ginger, peeled and thinly sliced
- ½ teaspoon Chinese five-spice powder

1. Place the chicken wings in a large saucepan with the onion, carrot, celery, garlic, ginger, and five-spice powder. (If using leftover roasted chicken, discard any skin. Cut the meat off the bones and refrigerate it in a covered bowl until ready to use.) Put the remaining carcass in the saucepan. Pour 8 cups cold water over the carcass, or a little more if needed to cover all the ingredients.
2. Cover the saucepan and bring the stock to a very gentle simmer (it should barely bubble). Simmer for at least 4 hours or up to 6 hours, if you have the time. Skim off any foam that rises to the surface and then top up the water, if needed.
3. Ladle the stock through a fine-mesh sieve into a large bowl or saucepan. Save any usable pieces of chicken meat from the bones and discard the rest of the ingredients. Serve the

broth with the reserved meat. You can use the broth right away as soup or cool completely before covering and storing in the fridge for a few days or the freezer for up to 2 months.

COOK'S TIP: You can prepare the stock in a slow cooker, letting it cook for several hours or overnight. Refer to the manufacturer's guidelines for correct quantities of water for the best results.

Chicken Noodles with Bok Choy

Bok choy is an excellent source of soluble fiber, as are mushrooms and green onions. We love this as a quick and easy light lunch—reminiscent of food from stalls in Southeast Asia.

245 calories: Serves 4

6⅓ cups Anti-Inflammatory Chinese-Style Chicken Broth (see page 196), or 6⅓ cups water with 1½ chicken or vegetable stock cubes dissolved in the water

5 ounces dried whole wheat or buckwheat noodles

5 ounces shiitake or another kind of mushrooms, sliced

4 to 7 ounces cooked chicken meat, chopped

4 green onions, trimmed and sliced

4 small heads bok choy, trimmed and thickly sliced

1 tablespoon sesame oil

1 long red chile, thinly sliced, or ½ teaspoon crushed red pepper flakes (optional)

1 to 1½ tablespoons dark soy sauce, to taste

1. Pour the stock into a large saucepan and add the noodles, mushrooms, cooked chicken, green onions, bok choy, oil, and chile, if using. Bring to a gentle simmer and cook for 3 minutes, or until the noodles are soft and the soup is hot.
2. Season with soy sauce to serve.

Tofu, Leek, and Kimchi Broth

A lovely, exotic Korean-inspired noodle soup that includes the benefits of fermented kimchi.

320 calories: Serves 2

1 tablespoon sesame oil
1 medium leek, trimmed and thinly sliced
2 garlic cloves, crushed
¼ cup kimchi
4 ounces dried whole wheat or soba noodles
2 cups hot vegetable or chicken stock (for fresh, see page
    196; or make it with boiling water and 1 stock cube)
5 ounces soft silken tofu, cut into roughly ¾-inch cubes
2 green onions, trimmed and thinly sliced
1 red chile, trimmed and thinly sliced
Handful of fresh cilantro, leaves coarsely chopped, to serve
Soy sauce, to serve

1. Heat the oil in a saucepan and fry the leek over medium heat for about 3 minutes, or until soft. Add the garlic and kimchi and cook for 1 minute longer, stirring.
2. Meanwhile, cook the noodles for 3 to 4 minutes, or according to the package instructions.
3. Pour the stock into the saucepan with the leek mixture and bring to a simmer. Add the tofu, green onions, and chile and cook for 3 minutes, stirring once.
4. Drain the noodles and divide them between two wide, shallow bowls. Spoon the broth on top, sprinkle with cilantro, and add soy sauce to taste.

Curried Lentil Soup with Turmeric

This rich, tasty soup offers gut-friendly fiber in the lentils along with the gorgeous golden glow of turmeric. Turmeric's anti-inflammatory health benefits are enhanced by combining it with coconut oil and black pepper.

580 calories: Serves 4

- 2 tablespoons olive or rapeseed oil
- 1 onion, finely chopped
- 1 tablespoon medium-strength curry powder
- 2 teaspoons ground turmeric
- 1½-ounce piece fresh ginger, peeled and finely chopped
- ¾ cup plus 2 tablespoons dried red split lentils (about 7 ounces)
- 1 (14-ounce) can full-fat organic coconut milk
- Juice of 1 lime or ½ lemon
- Flaked sea salt and freshly ground black pepper
- Handful of fresh cilantro, leaves coarsely chopped, to serve

For the crispy fried onions:
- 6 tablespoons olive or rapeseed oil
- 1 medium onion, sliced into thin rings

1. To make the soup, heat the oil in a large saucepan and gently fry the onion for about 5 minutes, or until softened, stirring regularly. Add the curry powder, turmeric, and ginger and cook for 1 minute longer, stirring.
2. Add the lentils, coconut milk, and 3 cups cold water. Bring to a gentle simmer, partially cover the pan with a lid, and cook for 20 to 25 minutes, or until the lentils are very soft,

stirring more frequently toward the end of the cooking time. Add a little extra water if needed to keep the lentils liquid and smooth.

3. Just before the soup is ready, prepare the crispy onions. Place the oil in a small saucepan over medium heat. Add the onion and fry for about 6 minutes, or until golden and crisp, stirring frequently. Take off the heat, remove the onion slices with a slotted spoon, and drain on paper towels.

4. Add the lime or lemon juice to the soup and season with salt and pepper to taste. Ladle into warmed bowls and top with golden onion rings and cilantro.

COOK'S TIP: To turn this recipe into a lentil dahl for a more substantial meal, cook with about 2½ cups water instead and top with halved hard-boiled eggs as well as the onion rings.

If making ahead, you may need to add extra water when reheating.

Baked Endive with Prosciutto

Endive root is one of the best sources of soluble fiber, supporting gut health and, by association, boosting mood. A light, flavorsome dish.

395 calories: Serves 2

1 tablespoon extra-virgin olive oil, plus extra for oiling

2 large heads endive (about 5 ounces each)

4 thin slices prosciutto or Parma ham (about 2 ounces total weight)

1½ ounces walnut halves, coarsely chopped

1 ounce Parmesan cheese, finely grated

2 teaspoons apple cider, white wine vinegar, or lemon juice

Freshly ground black pepper

1. Preheat the oven to 400°F. Lightly oil a small baking pan.
2. Trim a slim sliver off the base of each head of endive and cut the heads in half lengthwise. Put the endive cut side down on the baking pan and bake for 18 to 20 minutes, or until just tender.
3. Remove the pan from the oven and turn the endive heads over. Drizzle with the remaining oil. Drape the prosciutto over the top of each head of endive and sprinkle with the walnuts and Parmesan. Bake for 5 to 8 minutes longer, or until the walnuts are lightly toasted and the Parmesan is melted and beginning to brown.
4. Divide between two warmed plates, drizzle with the cider, season generously with pepper, and serve.

Dr. Tim's Hot-Smoked Salmon Salad

This delicious salad is so easy to make even Dr. Tim can do it. And it's super-healthy, too.

630 calories: Serves 2

5 ounces frozen edamame beans or baby fava beans

5 ounces broccoli, cut into small florets

5 ounces hot-smoked salmon or smoked mackerel fillets, flaked

2 ounces young spinach leaves (about 2 handfuls)

1 small ripe avocado, pitted, peeled, and sliced

¼ to ½ teaspoon crushed red pepper flakes, or 1 red chile, finely sliced (optional)

1 ounce flaked almonds, toasted

Juice of ½ lemon

2 tablespoons extra-virgin olive oil

Freshly ground black pepper

1. Fill a saucepan a third full with water and bring to a boil. Add the frozen edamame and the broccoli and return to a boil. Cook for 3 minutes, drain through a colander, and then rinse under running water until cool.
2. Place the vegetables in a large serving bowl and add the salmon or mackerel, spinach, avocado, and red pepper flakes, if using. Toss together lightly.
3. Sprinkle with the almonds, drizzle with the lemon juice and oil, season with pepper, and serve.

Griddled Eggplant with Feta and Pine Nuts

Luscious dark eggplant is the star of this dish, delivering plenty of fiber and antioxidants to help get your gut into top condition. Some research even suggests it helps lower blood sugars.

375 calories: Serves 2

½ teaspoon ground cumin
½ teaspoon ground coriander
Flaked sea salt and freshly ground black pepper
3 tablespoons extra-virgin olive oil
1 medium-large eggplant (about 11 ounces), trimmed and
    sliced lengthwise into 6 pieces
2 tablespoons pine nuts
4 ounces feta cheese, coarsely cubed
Large pinch of crushed red pepper flakes (optional)
About ½ ounce fresh cilantro, leaves coarsely chopped
Juice of ½ lemon

1. Preheat a large griddle pan until hot.
2. Mix the cumin and coriander in a small bowl with a good pinch of salt and lots of pepper. Add 2 tablespoons of the oil and stir well. Brush the eggplant slices on both sides with the seasoned oil.
3. Lay the eggplant slices on the hot griddle and cook without turning for 4 to 5 minutes, or until lightly browned. Turn the slices over and cook on the other side for 4 to 5 minutes, until browned and very tender. (Depending on the size of the griddle, you may have to cook the eggplant in batches. Keep the first batch warm in the oven while the second batch is cooking.)

4. While the eggplant is cooking, toast the pine nuts in a small dry frying pan over medium heat for 2 to 3 minutes, or until lightly browned, stirring regularly.

5. Divide the eggplant slices between two plates and top with the feta. Sprinkle with the pine nuts and red pepper flakes, if using. Scatter the cilantro over the eggplant and drizzle with the remaining 1 tablespoon oil and the lemon juice. Season with more pepper to serve.

COOK'S TIP: If you don't have a griddle pan, use a large nonstick frying pan. The eggplant slices won't have the same stripes as those cooked on a griddle or grill but will still taste delicious.

You can add a handful of baby spinach or arugula for extra greens, if you like.

Baked Sweet Potato with Smoked Mackerel

A tasty way to up your omega-3 intake.

630 calories: Serves 2

> 2 medium to large sweet potatoes (about 8 ounces each), scrubbed well
> 2 smoked mackerel fillets (about 3 ounces each), skin removed
> 10 cherry tomatoes, quartered
> ¼ cucumber, cut into roughly ¾-inch chunks
> 2 tablespoons good-quality mayonnaise
> 4 tablespoons full-fat live Greek yogurt
> Freshly ground black pepper
> 4 to 6 radishes, finely sliced, or pickled red onions (see page 263), to serve (optional)

1. Preheat the oven to 400°F. Place the sweet potatoes in a baking pan and prick each one a couple of times with a fork. Bake for 50 to 60 minutes, or until tender.
2. Just before the potatoes are ready, flake the mackerel fillets into a bowl and add the tomatoes, cucumber, mayonnaise, and yogurt. Season with pepper and mash together coarsely.
3. Divide the sweet potatoes between two plates and make a cross in the middle of each one. Open the potatoes by pushing the cross in on itself with your fingers. Stuff with the mackerel mixture, top with the radishes, if using, and serve.

COOK'S TIP: You could also cook the sweet potatoes in the microwave for about 8 minutes on high.

Pasta with Peas and Goat Cheese

Tasty and very easy to put together, this is made with ingredients you may already have stashed in the freezer, fridge, or cupboard, making it a great last-minute meal.

625 calories: Serves 2

3 ounces whole wheat pasta, such as penne or fusilli
2 tablespoons pine nuts
3 tablespoons extra-virgin olive oil
½ small onion, very finely chopped
1 small garlic clove, crushed
7 ounces frozen petits pois or English peas
Finely grated zest of ½ lemon and 1 tablespoon lemon juice
4 ounces goat cheese, rind removed
Flaked sea salt and freshly ground black pepper

1. Fill a saucepan halfway with water and bring to a boil. Add the pasta and cook for 10 to 12 minutes, or according to the package instructions, until just tender, stirring occasionally.
2. Meanwhile, lightly toast the pine nuts in a dry frying pan over medium heat for 2 to 3 minutes, or until lightly browned, stirring regularly. Transfer to a plate and set aside to cool.
3. Return the frying pan to the burner and add the oil and onion. Fry over low heat for 5 minutes, or until softened, stirring regularly. Add the garlic and peas and cook for a minute longer, stirring.
4. Drain the pasta and then return it to the saucepan. Add the fried onion and pea mixture and stir in the lemon zest and juice and the goat cheese. Toss over low heat until the cheese warms and begins to melt.
5. Season with salt and pepper and divide between two warmed bowls. Top with the toasted pine nuts to serve.

Hot-Smoked Salmon, Anchovies, Artichoke, and Broccoli Pasta

Omega-3 from the oily fish, lots of lovely fiber, as well as nutrient-rich fermented crème fraîche. All you need to soothe you to sleep.

610 calories: Serves 2

>   3 ounces whole wheat pasta, such as penne
>   1 small head broccoli (about 7 ounces), cut into small florets
>   4 anchovies in olive oil, from a jar or can, drained and chopped
>   3 tablespoons extra-virgin olive oil
>   1 small garlic clove, crushed
>   Leaves from 1 small sprig fresh rosemary, finely chopped
>   5 ounces hot-smoked salmon
>   3 ounces artichoke hearts, from a jar or can, marinated or in brine or oil, drained and cut into chunks
>   3 tablespoons full-fat crème fraîche
>   Small bunch of flat-leaf parsley, leaves coarsely chopped
>   Good pinch of crushed red pepper flakes (optional)
>   Freshly ground black pepper

1. Fill a saucepan halfway with water and bring to a boil. Add the pasta and cook for 10 minutes, or according to the package instructions, until just tender. Add the broccoli to the pasta for the last 3 minutes of the cooking time.
2. Drain the pasta and broccoli and return them to the pan. Cover loosely with a lid and set aside.
3. Place the anchovies, oil, garlic, and rosemary in a large non-stick frying pan and cook over low heat for about 1 minute, stirring until the anchovies soften and almost dissolve into the oil.

4. Add the salmon and artichoke hearts to the frying pan and gently heat through, 2 to 3 minutes, stirring gently until hot. Try not to let the salmon break up too much.
5. Add the pasta and broccoli to the frying pan, along with the crème fraîche, parsley, and red pepper flakes, if using. Season with black pepper and toss together gently. Divide between two warmed bowls to serve.

Super-Speedy Mushroom Risotto with Edamame

Risotto doesn't have to take hours. This lovely mushroom recipe can be made in 15 minutes and is packed with gut-friendly ingredients.

410 calories: Serves 2

½ ounce dried porcini mushrooms (or any dried wild mushrooms)

2 tablespoons olive oil

1 small onion, finely chopped

4 ounces fresh cremini mushrooms, sliced

1 garlic clove, crushed

1 cup plus 2 tablespoons cooked brown rice, or mixed brown and wild rice

⅓ cup frozen edamame beans (about 3 ounces)

1 tablespoon chia seeds

Pinch of dried thyme

¾ cup plus 2 tablespoons chicken or vegetable stock (fresh or made with ½ stock cube)

1 ounce Parmesan cheese, finely grated

Flaked sea salt and freshly ground black pepper

Pinch of crushed red pepper flakes (optional)

1. Place the dried mushrooms in a heatproof bowl and cover with ⅔ cup just-boiled water. Let the mushrooms soak for about 10 minutes to soften. Drain them in a sieve, reserving the soaking liquor.
2. Meanwhile, heat the oil in a large nonstick saucepan and gently fry the onion for 3 to 4 minutes, or until softened, stirring regularly. Add the cremini mushrooms, increase

the heat, and cook for 2 to 3 minutes more, or until the mushrooms are lightly browned.

3. Add the garlic and cook for a few seconds before stirring in the soaked and drained porcini (coarsely chopping any particularly large ones), along with the reserved soaking liquor.

4. Add the rice, edamame beans, chia seeds, and thyme and stir for about 1 minute.

5. Pour the stock into the pan and bring to a gentle simmer. Cook for about 3 minutes, stirring regularly.

6. Stir half the Parmesan into the risotto and season with salt, black pepper, and red pepper flakes, if using. Spoon into two warmed shallow bowls and sprinkle with the remaining Parmesan to serve.

COOK'S TIP: You could use precooked brown/wild rice from a package, but make sure it doesn't contain any flavorings or added ingredients (except for a little oil, which is used to keep the grains separate). If you are cooking rice especially for this recipe, you will need about ⅓ cup of dried mixed brown and wild rice to yield about 1 cup plus 2 tablespoons of cooked rice.

If you can't get hold of edamame beans, use frozen peas or fava beans instead.

Sardines on Seeded Sourdough

Brainpower on toast, with omega-3 to oil those cogs.

310 calories: Serves 2

    1 (4-ounce) can sardines, packed in olive oil
    Juice of ½ small lemon (about 1 tablespoon)
    Flaked sea salt and freshly ground black pepper
    2 thin slices seeded sourdough bread (about 1½ ounces
       each)
    ¼ small red onion, very thinly sliced
    1 teaspoon drained baby capers (optional)
    1 tablespoon extra-virgin olive oil

1. In a mixing bowl, coarsely mash the sardines in their oil with the lemon juice. Season with a little salt and plenty of pepper.
2. Toast the bread and divide between two plates—or make this for one if you are especially hungry.
3. Spread the mashed sardines on the hot toast and top with the onion, capers (if using), and a drizzle of oil. Serve immediately, before the toast has time to cool.

ON THE MOVE

It's challenging to eat healthy meals when you are out and about, particularly if you are a shift worker looking for food outside normal working hours. Being prepared will help ensure that you have the right foods on hand and are not dependent on unhealthy snacks from vending machines or gas stations. There are some great recipes here that are both portable and nutritious; and do try the ones with seaweed, the new omega-3-rich superfood.

Celery with Blue Cheese Dip

Celery may contain lots of water, but it also has plenty of gut-friendly fiber, both soluble and insoluble, while blue cheese adds a rich, tangy flavor to the dip, as well as a dose of healthy microbes. This is also yummy served on whole-grain seeded crackers.

225 calories: Serves 2

> ¼ cup crumbled blue cheese, such as Roquefort (about 2 ounces)
> ¼ cup full-fat live Greek yogurt
> ¼ cup full-fat crème fraîche
> Freshly ground black pepper
> Celery stalks or other vegetable crudités, to serve

1. Place the cheese, yogurt, and crème fraîche in a bowl and mash with a fork until thoroughly combined. Season to taste with pepper.
2. Spoon into a small dish, or lidded container if taking as a packed lunch. Serve with celery stalks for dipping.

Cashew, Ricotta, and Red Pepper Dip

A tangy and tasty dip with extra nutrients and creaminess from the cashews.

230 calories: Serves 4

4 ounces plain cashews (not roasted)
1 small garlic clove, coarsely chopped
3 ounces jarred roasted red peppers, drained
½ cup ricotta cheese (about 4 ounces)
4 tablespoons extra-virgin olive oil, plus more for drizzling
Finely grated zest of ½ small lemon
Flaked sea salt and freshly ground black pepper
Pinch of paprika, to serve (optional)

1. Put the nuts in a bowl and add just enough cold water to cover. Refrigerate for 4 to 6 hours to soak and soften.
2. Drain the nuts and transfer to a food processor fitted with the metal blade. Add the garlic, peppers, cheese, oil, and lemon zest. Blitz until thoroughly combined. Season with salt and pepper.
3. Spoon into a serving dish, drizzle with a little oil, and sprinkle with paprika, if using. Serve with lots of vegetable sticks or sourdough flatbread.

COOK'S TIP: If you don't have a food processor, place the soaked cashews and the remaining ingredients in a bowl or large pitcher and mix with an immersion blender instead.

You could use home-roasted red bell peppers instead of the jarred kind, if you like.

Shrimp and Red Cabbage Slaw on Sourdough

Shrimp and slaw on sourdough bread is an all-around, gut-friendly winner—easy to assemble and tasty, too. Sourdough is one of the most healthful breads available, made using a slow process of fermentation, which makes it easier to digest and less likely to cause a spike in blood sugar. Michael and I are big fans, and love the flavor and firm texture, too.

270 calories: Serves 2

For the dressing:
2 tablespoons extra-virgin olive oil
1 teaspoon fresh lemon juice, plus extra to serve
½ teaspoon Dijon mustard
½ teaspoon honey (optional)

For the slaw:
½ cup loosely packed red cabbage, trimmed and very finely
    sliced (about 4 ounces)
1 small carrot, scrubbed well and coarsely grated
1 tablespoon mixed seeds, such as pumpkin, sunflower,
    sesame, and flax

2 slices sourdough bread
Small handful of fresh watercress or mixed baby salad leaves
    (about 1 ounce)
3 ounces cooked and peeled shrimp
Freshly ground black pepper

1. To make the dressing, whisk together the oil, lemon juice, mustard, and honey, if using, in a large bowl.

216

2. To make the slaw, add the cabbage, carrot, and mixed seeds to the bowl with the dressing and toss together.

3. Divide the sourdough between two plates and top with the watercress. Place some cabbage slaw on top and then scatter the shrimp on the salad. Squeeze a little extra lemon juice over the salad and season with pepper to serve.

COOK'S TIP: You can use frozen cooked shrimp. Let them thaw and drain before using.

Smoked Mackerel, Beets, and Tahini Pita

Juicy, filling, and full of flavor, this is a nutritious instant meal—and portable, too. Oily fish has been found not only to improve sleep, but also to improve function during the day.

635 calories: Serves 1

For the dressing:

2 tablespoons full-fat live Greek yogurt

1 tablespoon tahini

1 tablespoon extra-virgin olive oil

1 to 2 teaspoons fresh lemon juice

Flaked sea salt and freshly ground black pepper

For the pita:

1 piece brown pita bread

Small handful of fresh watercress or mixed baby salad leaves (about ½ ounce)

1 (3-ounce) fillet smoked mackerel

1 cooked beet, peeled and sliced

Flaked sea salt and freshly ground black pepper

1. To make the dressing, thoroughly mix the yogurt, tahini, oil, and lemon juice with ¼ cup cold water in a small bowl. Season with salt and pepper.
2. To make the pita, warm the pita, if possible, and then split it open. Fill the pocket with the watercress. Flake the mackerel and put the flakes into the pita pocket along with the beets. Season with plenty of pepper and drizzle with a little of the dressing.

COOK'S TIP: The rest of the dressing can be refrigerated for up to 2 days. Use on salads or as a dip.

Black Bean Salad with Lime and Avocado

A tangy and filling salad. The tasty black beans are not only a boost of fiber but also add a generous number of anti-inflammatory phytonutrients and B vitamins, including folate.

315 calories: Serves 2

1 (14-ounce) can black beans, rinsed and drained
½ teaspoon flaked sea salt
4 ounces sugar snap peas or slender green beans, trimmed
1 small ripe but firm avocado, peeled, pitted, and sliced
2 green onions, trimmed and finely sliced
1 red chile, finely chopped, or ½ teaspoon crushed red
  pepper flakes
About ½ ounce fresh cilantro, leaves coarsely chopped

For the dressing:
2 tablespoons extra-virgin olive oil
1½ tablespoons fresh lime juice
1 tablespoon sesame seeds (about ½ ounce), toasted
Freshly ground black pepper

1. Place the black beans in a serving bowl and toss with the salt. Set aside while preparing the rest of the recipe.
2. Fill a saucepan a third full with water and bring to a boil. Add the peas and cook for 2 minutes, until tender but still crisp. (If using green beans, cook for 3 minutes.) Drain the peas and then rinse under running water until cold. Drain again.
3. Put the peas into the bowl with the black beans and add the avocado, onions, chile, and cilantro.

4. To make the dressing, whisk the oil, lime juice, and sesame seeds together in a small bowl and season with lots of pepper. Drizzle the dressing over the bean salad to serve.

COOK'S TIP: To toast the sesame seeds, sprinkle in a dry frying pan and place over medium heat. Stir constantly for 1 to 2 minutes, until lightly browned. Remove from the heat and put in a small bowl so they don't continue to brown.

Nori Rolled Crab

The extra omega-3 in the seaweed, along with the crab, make this nori roll both super-tasty and super-healthy for your gut.

420 calories: Serves 2

½ medium avocado, peeled and pitted
4 ounces cooked crabmeat, fresh or canned, drained
1 tablespoon fresh lime juice
Freshly ground black pepper
2 sheets dried nori (each about 7 inches square)
1 cup cooked and cooled brown rice

For the dip:
2 tablespoons dark soy sauce
¼ teaspoon crushed dried chili flakes or a dash of sriracha chili sauce
1 teaspoon sesame oil

1. Place the avocado on a plate and mash well with a fork.
2. Put the crab in a bowl, add the lime juice, and coarsely mash them together. Season with a good grinding of pepper.
3. Place a nori sheet on a cutting board, shiny side down, and spread half of the avocado evenly over the bottom half of the sheet.
4. Top the avocado with half of the rice and press down lightly with the back of a spoon. Spread half of the crab mixture in a horizontal line across the center of the rice.
5. Using both hands, roll the nori from the bottom up, around

the filling, pressing firmly. Seal the roll by brushing it with a little water and pressing the edges together. Trim the edges, and then cut the roll into 6 pieces. Repeat with the remaining sheet of nori and the rest of the filling to make 6 more pieces, for a total of 12.

6. For the dip, mix the soy sauce, chili flakes, and sesame oil in a small bowl and serve alongside the nori rolls as a dip.

COOK'S TIP: You can use leftover rice for this recipe—as long as it was cooked and then cooled right away—or boil ¼ cup of brown rice according to the package instructions until tender. Rinse under cold running water and drain well. You should find that ¼ cup of dried rice will give you about ½ cup plus 2 tablespoons of cooked rice.

Nori Chili Popcorn

These are heathy nibbles with an omega-3 boost and an ir-resistible umami flavor.

125 calories: Serves 2

½ sheet nori, torn into roughly 1¼-inch strips
½ teaspoon crushed red pepper flakes
¼ teaspoon flaked sea salt

3 tablespoons extra-virgin olive oil
2 tablespoons popping corn

1. To make the nori chili seasoning, put the nori, red pepper flakes, and salt in a deep bowl or pitcher and blitz it with an immersion blender until very finely chopped but not too powdery.
2. Put 1 teaspoon of the oil in a large saucepan. When it is warm, put the corn in the saucepan and stir to coat with oil. Cover the pan with a tight-fitting lid and set over medium heat. As soon as you hear the corn popping, gently shake the pan forward and backward. Don't be tempted to take off the lid—and cook for about 4 minutes longer or until the corn stops popping, shaking the pan regularly while holding the lid firmly in place.
3. Remove the pan from the heat and drizzle with the remaining oil. Toss well, then add the nori seasoning and toss again. Pour the popcorn into a large bowl to serve, discarding any kernels that haven't popped.

Tabbouleh with Goat Cheese

Bulgur wheat is a delicious nutty grain, often used to make tabbouleh with chopped salad and herbs. In this recipe it's topped with cheese and a tangy garlic dressing, to make an ideal light and portable meal. Add leftover cold meats, falafel, or nuts to make a more filling dish.

365 calories: Serves 4

1 cup bulgur wheat (ideally whole grain)
½ red onion, very finely sliced
1 ounce fresh mint, leaves coarsely chopped
1 ounce flat-leaf parsley, leaves coarsely chopped
2 ounces young spinach leaves
Flaked sea salt and freshly ground black pepper
4 ounces goat cheese, cut into small chunks
2 tablespoons mixed seeds, such as pumpkin, sunflower, sesame, and flax

For the dressing:
2 garlic cloves, crushed
2 tablespoons fresh lemon juice
4 tablespoons extra-virgin olive oil

1. Fill a medium saucepan halfway with water and bring to a boil over medium-high heat. Add the bulgur wheat, return to a boil, lower the heat, and rapidly simmer for about 10 minutes, or until almost tender, stirring occasionally. Drain in a sieve and rinse well under running water until cool. Drain thoroughly and transfer to a large serving bowl.
2. Add the onion, mint, parsley, and spinach to the bowl and

season with a good pinch of salt and plenty of pepper. Toss well to mix.

3. To make the dressing, whisk the garlic, lemon juice, and oil in a small bowl until well combined. Pour over the salad and toss lightly.

4. Scatter the goat cheese and seeds over the top of the salad to serve.

COOK'S TIP: You can use any cheese you like for this recipe, or you can leave it out altogether. Serve the tabbouleh as an accompaniment to grilled meat, fish, or roasted wedges of squash or beets.

MAIN MEALS

These are more substantial dishes, still based on a Mediterranean style of eating, with extra beans, lentils, and other good sources of fiber to help boost your microbiome and enhance your sleep.

Trout with Celeriac Mash

You just can't beat fresh fish with lemon and a crunchy, nutty topping. This dish comes with a creamy celeriac mash; we recommend minimal peeling so that you get the benefit of all those extra nutrients found in the vegetable's skin. Serve with some freshly cooked green vegetables.

555 calories: Serves 2

    2 (10-ounce) rainbow trout or fresh mackerel, cleaned
    1 small lemon, thinly sliced
    2 tablespoons extra-virgin olive oil
    Freshly ground black pepper
    1½ ounces blanched hazelnuts, coarsely chopped
    11 ounces celeriac, scrubbed well, gnarled bits removed, cut
        into roughly ¾-inch chunks
    ¼ cup full-fat crème fraîche
    Flaked sea salt

1. Preheat the oven to 400°F. Line a baking pan with parchment paper.
2. Place the trout or mackerel in the pan and insert half of the lemon slices into each fish. Drizzle ½ tablespoon of oil over each fish and season with pepper. Bake, uncovered, for 15 minutes. Remove from the oven and scatter with the hazelnuts. Bake for another 5 minutes, or until the hazelnuts are lightly toasted.
3. Meanwhile, put the celeriac pieces in a saucepan and cover with cold water. Bring to a boil and cook for 10 minutes, or until very soft. Drain and then return the pieces to the pan.
4. Add the crème fraîche, a little salt, and lots of pepper to the pan and blitz with an immersion blender until the mash is smooth.
5. Divide the mash between two warmed plates, put the fish on the plates, and spoon any loose hazelnuts over the top. Drizzle with the remaining oil and serve.

Beef and Jerusalem Artichoke Casserole

The knobby Jerusalem artichoke offers masses of gut-friendly insoluble and soluble fiber, and has a lovely, nutty, and slightly sweet taste. Serve with some celeriac mash and lots of green leafy vegetables. Celeriac is also called celery root.

320 calories: Serves 4

> 1 pound braising or stewing beef steak, trimmed and cut into roughly 1-inch chunks
> Flaked sea salt and freshly ground black pepper
> 3 tablespoons olive oil
> 1 onion, thinly sliced
> 2 celery stalks, cut into roughly ½-inch slices
> 3 medium carrots, trimmed, scrubbed well, and cut into roughly ¾-inch chunks
> 11 ounces Jerusalem artichokes, scrubbed well and cut into roughly ¾-inch chunks
> 1 beef stock cube
> 2 tablespoons tomato puree
> 1 teaspoon dried mixed herbs

1. Preheat the oven to 325°F. Season the beef all over with salt and pepper.
2. Heat 1 tablespoon of the oil in a large nonstick frying pan and fry the beef in two batches over medium-high heat for 2 to 3 minutes, or until browned, adding a little more oil if needed to prevent sticking. Transfer to a flameproof casserole dish.
3. Add the remaining oil to the frying pan and cook the onion, celery, carrots, and artichokes for 6 to 8 minutes, or until

lightly browned, stirring regularly. Add to the dish with the beef.

4. Leaving the pan on the heat, pour ½ cup of just-boiled water into the pan and scrape the base of the pan with a wooden spoon to dissolve the bits stuck to the bottom. Pour the liquid into the casserole with the beef and vegetables. Add the stock cube, tomato puree, and 1½ cups more of just-boiled water. Sprinkle with the herbs and bring to a simmer, stirring occasionally. Cover with a lid and cook in the oven for about 2 hours, or until the beef is very tender.

Cheesy Celeriac and Anchovy Bake

The taste of the celeriac is wonderfully enhanced by the anchovies, and the health benefits of the dish are also boosted by the oily fish. Serve this with spring greens, kale, or cavolo nero, a green that also is called Italian kale or Tuscan kale and has dark leaves.

420 calories: Serves 2

About 14 ounces celeriac, scrubbed and gnarled bits
    removed
1 small onion, sliced
1 tablespoon olive oil
Flaked sea salt and freshly ground black pepper
8 anchovies in olive oil (from a jar or can), drained
1 ounce Gouda or cheddar cheese (or a mixture), coarsely
    grated
5 tablespoons full-fat crème fraîche
4 tablespoons whole milk
1 ounce Parmesan cheese, finely grated
1 teaspoon finely chopped fresh rosemary

1. Preheat the oven to 400°F. Cut the celeriac in half and then into very thin slices.
2. Scatter the onion over the base of an ovenproof dish, drizzle it with the oil, and season with salt and pepper. Arrange the slices of celeriac on top of the onion. The slices should be 3 to 4 layers deep. Lay the anchovies on top of the celeriac and then scatter with the Gouda.
3. Mix the crème fraîche with the milk until it is a pouring consistency and then spoon it over the celeriac. Season with more pepper. Cover the dish loosely with foil and bake in

the center of the oven for 55 to 60 minutes, or until the celeriac is just tender.

4. Remove from the oven and discard the foil. Scatter the Parmesan and rosemary over the top and then return the dish to the oven for about 10 minutes, or until pale golden brown.

Chicken and Vegetable Tray Bake

An easy, all-in-one, gut-friendly dish. Serve with a crisp green salad.

530 calories (without chorizo): Serves 4

2 medium red onions, each cut into 8 wedges
11 to 12 ounces sweet potatoes, scrubbed well and cut into roughly 1¼-inch chunks
11 to 12 ounces Jerusalem artichokes (or 5 to 6 ounces sweet potato and an extra bell pepper), scrubbed well and cut into 1½-inch chunks
2 bell peppers (1 red and 1 yellow, if possible), seeded and cut into roughly 1¼-inch chunks
2 ounces chorizo, cut into roughly ¾-inch chunks (optional)
4 large tomatoes (about 1 pound), quartered
4 tablespoons extra-virgin olive oil
4 bone-in, skin-on free-range chicken thighs (about 1⅔ pounds total weight)
Flaked sea salt and freshly ground black pepper

1. Preheat the oven to 400°F.
2. Place the onions, sweet potatoes, artichokes, peppers, chorizo (if using), and tomatoes in a large, deep roasting pan. Drizzle with 1 tablespoon of the oil and toss together lightly.
3. Nestle the chicken thighs among the vegetables, skin side up. Drizzle with the remaining 3 tablespoons oil and season with salt and pepper. Roast for 45 to 55 minutes, or until the chicken is thoroughly cooked and the vegetables are lightly browned.

Chicken Tagine with Chickpeas and Dates

For this fabulous tagine, simply throw ingredients in a casserole and let the meat soften and the flavors develop. Serve with quinoa or bulgur wheat and a generous pile of green vegetables.

540 calories: Serves 4

4 tablespoons extra-virgin olive oil

1 large onion, thinly sliced

2 garlic cloves, crushed

2 tablespoons harissa paste

8 boneless, skinless free-range chicken thighs (about 1¾ pounds total weight)

2 (14-ounce) cans chopped tomatoes

1 (14-ounce) can organic chickpeas

8 soft dates, pitted and cut into thick slices

1½ ounces blanched almonds (optional)

1 chicken stock cube

Flaked sea salt and freshly ground black pepper

1 ounce fresh cilantro, stalks finely chopped and leaves coarsely chopped

1. Preheat the oven to 400°F.
2. Heat the oil in a flameproof casserole dish over medium heat and add the onion, garlic, and harissa. Stir for 30 seconds to 1 minute on the heat. Add the chicken, tomatoes, chickpeas and their liquid, dates, almonds, stock cube, and ½ cup water. Season with a pinch of salt and lots of pepper. Bring to a gentle simmer, stirring occasionally.

3. Add the cilantro stalks and then cover the casserole with a lid and cook in the oven for about 1 hour, stirring occasionally, or until the chicken is very tender and the sauce has thickened. Stir half the cilantro leaves into the tagine, and then sprinkle with the rest just before serving.

Meatballs with Feta and Eggplant

These meatballs are beautifully complemented by the eggplant, with its antioxidants for reducing inflammation in the body and brain. Serve with leafy greens or a salad, and for a more substantial meal, with small portions of brown rice or bulgur wheat.

420 calories: Serves 4

4 tablespoons olive oil

20 small premade lamb meatballs (about 1 pound total)

1 red onion, finely chopped

1 eggplant (about 11 ounces), cut into roughly ¾-inch chunks

1 (14-ounce) can chopped tomatoes

½ cup red wine (about 1 small wineglass)

1 teaspoon dried oregano

½ teaspoon crushed red pepper flakes

Flaked sea salt and freshly ground black pepper

4 ounces feta cheese

1. Heat 1 tablespoon of the oil in a large, deep nonstick frying pan set over medium heat. Add the meatballs and fry for 8 to 10 minutes, or until lightly browned on all sides, turning regularly. Transfer the meatballs to a plate and spoon off the excess fat in the pan. Return the empty pan to the heat.
2. Add the remaining 3 tablespoons oil to the pan and fry the onion and eggplant for 6 to 8 minutes, or until the onion is softened and the eggplant is lightly browned, stirring regularly.
3. Add the tomatoes to the pan and then add the red wine, oregano, red pepper flakes, and ¾ cup cold water. Season with salt and black pepper and bring to a gentle simmer.

Return the meatballs to the pan and cook in the sauce for about 15 minutes, stirring regularly. Add a splash more water if needed.

4. Finally, crumble the feta over the top and gently simmer without stirring for another 3 to 4 minutes, or until the cheese is hot and has melted slightly into the sauce.

COOK'S TIP: You can make this recipe using beef meatballs, veggie meatballs, or prepared falafel. If using falafel, serve 2 to 3 per person and don't cook them ahead. Instead, add them to the sauce after it has been simmering for 10 minutes.

Mushroom, Chickpea, and Kale Curry

A super-healthy curry—easy, creamy, and full of flavor. Serve with brown rice and a tomato and onion salad.

580 calories: Serves 3

3 tablespoons extra-virgin olive oil

1 onion, finely chopped

7 ounces small cremini mushrooms, quartered

2 garlic cloves, crushed

1 ounce fresh ginger, peeled and finely chopped

1 to 1½ tablespoons medium-strength curry powder, to taste

1 (14-ounce) can full-fat organic coconut milk

5 ounces curly kale, thickly shredded and tough stalks discarded

1 (14-ounce) can organic chickpeas, not drained, or other legumes

1½ ounces cashews (not roasted), coarsely chopped

Flaked sea salt and freshly ground black pepper

Squeeze of lemon or lime juice (optional)

1. Heat the oil in a large wide-based saucepan or shallow casserole and gently fry the onion for 2 to 3 minutes, stirring regularly. Increase the heat to medium-high, add the mushrooms, and cook for 2 to 3 minutes longer or until lightly browned, stirring.
2. Add the garlic, ginger, and curry powder and cook for another 30 seconds, stirring constantly. Add the coconut milk and stir in the kale, chickpeas and their liquid, and cashews. Bring to a simmer, cover with a lid, and cook for 10 minutes, or until the kale is tender, stirring occasionally.

3. Season to taste with salt and pepper, adding a squeeze of lemon or lime juice, if using, to serve.

COOK'S TIP: Use organic chickpeas packed in water. You can add them right from the can. If you can't find organic chickpeas, rinse and drain any you buy. If making the curry with regular chickpeas, add an extra ½ cup of water to the curry.

Ratatouille with White Beans

This works well as a main meal or a side—a glorious, soluble-fiber-rich ratatouille. It is also rich in anti-inflammatory olive oil. Enjoy the ratatouille on its own or served alongside fried halloumi, cooked meats, or fish.

545 calories: Serves 2 as a main or 4 as a side dish

5 tablespoons extra-virgin olive oil

2 medium onions, finely chopped

7 to 8 ounces cremini mushrooms, sliced

1 eggplant (about 12 ounces), cut into roughly ¾-inch chunks

4 garlic cloves, crushed

1 teaspoon dried oregano

1 (14-ounce) can organic white beans, such as cannellini, haricot, or butter beans (do not drain)

1 (14-ounce) can chopped tomatoes

1 vegetable stock cube

Flaked sea salt and freshly ground black pepper

1. Heat 4 tablespoons of the oil in a wide-based saucepan or shallow casserole and gently fry the onions for 6 to 8 minutes, or until softened, stirring regularly.
2. Add the mushrooms and eggplant, increase the heat, and fry until the eggplant is lightly browned, stirring constantly.
3. Add the garlic and oregano and cook for a few seconds more, stirring.
4. Add the beans and their liquid and then add the tomatoes. Crumble the stock cube over the ratatouille and then stir 1 cup plus 2 tablespoons water into the pan. Bring to a gentle simmer, cover with a lid, and cook for 20 minutes.

Remove the lid and cook for another 10 minutes, or until the vegetables are soft and the tomatoes have reduced until lightly thickened.

5. Season to taste with salt and pepper and drizzle with the remaining 1 tablespoon oil to serve.

Shepherd's Pie with Parsnip and White Bean Mash

There is lots of glorious fiber in this shepherd's pie, thanks to all the vegetables, lentils, and beans. Good, sleep-inducing comfort food. Serve with plenty of freshly cooked green leafy vegetables.

355 calories: Serves 6

- 1 tablespoon olive oil
- 14 ounces ground lamb
- 1 onion, finely chopped
- 1 celery stalk, trimmed and thinly sliced
- 1 carrot, scrubbed, trimmed, and cut into roughly ¼-inch chunks
- 4 ounces mushrooms, sliced
- 4 ounces dried green lentils
- 2 cups lamb or vegetable stock (fresh or made with 1 stock cube)
- 2 tablespoons tomato puree
- 1 bay leaf
- ½ teaspoon dried thyme

For the topping:
- 11 ounces parsnips, trimmed, washed well, and cut into roughly ¾-inch chunks
- 1 (14-ounce) can cannellini beans, drained and rinsed
- 4 tablespoons full-fat crème fraîche
- 2 to 3 tablespoons whole milk
- Flaked sea salt and freshly ground black pepper

1. Preheat the oven to 350°F.
2. Heat the oil in a flameproof casserole (it will need to hold about 12 cups or 3 quarts of food) and gently fry the lamb, onion, celery, carrot, and mushrooms for 10 minutes, or until lightly browned, stirring regularly to break up the ground meat.
3. Add the lentils, stock, tomato puree, bay leaf, and thyme and bring to a simmer. Cover with a lid, transfer to the oven, and cook for about 1 hour, or until the lentils are tender and the sauce is thick.
4. To make the topping, put the parsnips in a large saucepan and cover with cold water. Bring to a boil and cook for 15 to 20 minutes, or until tender. Add the drained beans and simmer for another 2 minutes.
5. Drain the parsnips and beans and return them to the pan. Add the crème fraîche and milk, season with salt and pepper, and mash until well mixed
6. Remove the lamb mixture from the oven and stir well. Spoon the parsnip mixture on top, spreading it as evenly as possible. Return to the oven, uncovered, and cook for 15 minutes longer, or until the parsnip mash is hot.

COOK'S TIP: You could transfer the cooked lamb to a warmed flameproof pie dish before topping and heating.

Smokey Fish Gratin

A simple, creamy fish pie with lots of flavor and a crispy topping—comfort food rich in omega-3.

570 calories: Serves 2

> 1 teaspoon olive oil, for greasing
> 10 ounces mixed pieces of fish, such as salmon, cod, and smoked haddock, cut into ¾-inch chunks
> 3 ounces frozen peas (about ⅓ cup)
> Freshly ground black pepper
> 4 ounces full-fat crème fraîche
> 1 ounce whole grain sourdough bread crumbs (about 1 tablespoon)
> 2 ounces cheddar cheese, coarsely grated
> Small handful of chopped flat-leaf parsley (optional)

1. Preheat the oven to 400°F. Grease a shallow ovenproof dish with the oil. (The dish will need to be large enough to hold about 2 cups of food.)
2. Place the fish and peas in the dish, season with a little pepper, and toss lightly. Cover with foil and bake for 10 minutes.
3. Remove the dish from the oven and gently stir in the crème fraîche.
4. Mix the bread crumbs and cheese together in a bowl and sprinkle over the top of the fish. Garnish with the parsley, if using. Return to the oven for another 10 to 15 minutes, or until the crumbs are golden and the filling is bubbling hot.

COOK'S TIP: If you can't get hold of a whole grain sourdough loaf, use any whole grain loaf or sourdough bread to make the crumbs.

Turkey and Lentil Bolognese

Enjoy this gut-friendly Bolognese with freshly cooked zucchini or small portions of whole wheat spaghetti and lots of grated Parmesan.

415 calories: Serves 4

    4 tablespoons olive oil
    1 large onion, finely chopped
    5 ounces cremini mushrooms, sliced
    11 ounces ground turkey
    2 garlic cloves, crushed
    4 ounces dried red split lentils
    2 (14-ounce) cans chopped tomatoes
    $2/3$ cup red wine
    1 chicken stock cube
    1½ teaspoons dried oregano
    1 to 2 bay leaves (optional)
    Flaked sea salt and freshly ground black pepper

1. Heat the oil in a large nonstick saucepan and gently fry the onion, mushrooms, and turkey for 5 to 7 minutes, or until the onion is softened and the turkey is lightly colored, stirring regularly. Add the garlic and cook for a few seconds more, stirring.
2. Add the lentils to the pan, along with the tomatoes, red wine, stock cube, oregano, and bay leaves, if using. Add ¾ cup water and bring to a simmer. Reduce the heat and simmer gently for about 30 minutes, or until the lentils soften, the turkey is tender, and the sauce is thick, stir-

ring occasionally. Add a little extra water if the sauce gets too thick.

3. Season to taste with salt and pepper to serve.

COOK'S TIP: There are plenty of gluten-free, high-fiber pasta alternatives, including those made using buckwheat, pea, or lentil flour.

Nori-Flavored Warm Salmon and Squash Salad

A delicious and colorful poke bowl, with Asian flavors and lots of texture. Full of anti-inflammatory omega-3 from the oily fish with an extra boost from the seaweed. All this combined with gut-friendly fiber should enhance sleep.

590 calories: Serves 2

For the nori seasoning:
1 sheet nori, cut into wide strips
½ teaspoon crushed red pepper flakes
2 tablespoons sesame seeds
Good pinch of ground ginger (optional)
Flaked sea salt

For the salad:
11 ounces butternut squash, peeled and cut into roughly
    ¾-inch chunks
1 tablespoon extra-virgin olive oil
2 salmon fillets (about 4 ounces each), preferably with skin on
Freshly ground black pepper
4 ounces frozen edamame beans or peas
2 large handfuls young spinach leaves (about 3 ounces)
4 radishes, sliced, or 1 carrot, scrubbed, trimmed, and
    coarsely grated
2 green onions, trimmed and finely sliced
2 tablespoons dark soy sauce
2 teaspoons sesame oil

1. To make the seasoning, put the nori, red pepper flakes, sesame seeds, ginger (if using), and a good pinch of salt in

a blender and blitz into tiny pieces, without allowing it to become a powder. Set aside.

2. Preheat the oven to 400°F. Line a baking sheet with parchment paper. To make the salad, toss the squash with the olive oil and then spread it on the prepared baking sheet. Bake for 20 minutes.

3. Remove the baking sheet from the oven and turn the squash pieces over. Add the salmon to the baking sheet, skin side down. Season with black pepper and bake for 10 to 12 minutes, or until the salmon is just cooked and the squash is tender.

4. Cook the edamame beans in a pan of boiling water for 2 minutes. Drain and divide them between two wide bowls. Divide the spinach leaves, radishes, and green onions between the same two bowls.

5. Top the bowls with the cooked squash and salmon (you should be able to lift the salmon off its skin once it is cooked).

6. Mix the soy sauce and sesame oil in a small bowl, and drizzle the mixture over the salad. Sprinkle a little of the reserved nori seasoning over both bowls to serve.

COOK'S TIP: Keep the remaining seasoning in a sealed jar and use for other salads or to sprinkle on soups, stews, and stir-fries.

For a vegetarian alternative, omit the salmon and cook an extra 4 ounces of squash. Top each salad with toasted flaked almonds or hazelnuts for added protein.

VEGETABLE SIDES

Go to sleep on your vegetables. The greener, more color-ful and diverse the better. Vegetables, along with beans, lentils, and whole grains, are a key part of the Med-style diet—both for their taste and the huge variety of benefi-cial nutrients, vitamins, and protein they deliver. Aim to eat 30 different varieties of vegetables, fruit, and legumes a week—this is easier than you think, although build up to it gradually, as your gut may not be ready for a sudden switch.

Baked Turmeric-Spiced Fennel and Onion

Combining soluble fiber and the anti-inflammatory effects of turmeric, this juicy, delicately flavored dish goes with almost anything.

160 calories: Serves 2

2 tablespoons olive oil
1 teaspoon ground turmeric
1½ tablespoons fresh lemon juice
Flaked sea salt
2 small fennel bulbs (about 8 ounces each), trimmed and
    sliced lengthwise into 8 pieces
1 small onion, quartered
Freshly ground black pepper
Chopped fresh cilantro, to serve (optional)

1. Preheat the oven to 400°F.
2. Mix the oil with the turmeric and lemon juice in a large bowl. Add a good pinch of salt and whisk together just to mix.
3. Add the fennel and onion to the bowl and toss well. Season with pepper. Spread the vegetables over a large baking pan in a single layer. Bake for about 25 minutes, or until the fennel is softened and lightly browned.
4. Scatter with the cilantro, if using, to serve.

Braised Red Cabbage with Walnuts and Apple

Wonderful winter food. Serve with cold, cooked meats or crumble microbe-rich blue cheese on top.

215 calories: Serves 4

- 1 small red cabbage (about 1 pound), quartered, cored, and thinly sliced
- 2 tablespoons extra-virgin olive oil
- ½ teaspoon flaked sea salt
- 2 star anise (optional)
- 1 cinnamon stick or 1 teaspoon ground cinnamon
- 1 cooking apple (such as Granny Smith, about 8 ounces), unpeeled, quartered, cored, and cut into roughly 1½-inch chunks
- 2 ounces soft pitted dates, thinly sliced
- 2 ounces walnut halves, very coarsely chopped
- 1½ tablespoons red wine vinegar or apple cider vinegar

1. Place the cabbage in a large saucepan with ⅔ cup cold water. Stir in the oil, salt, anise (if using), and cinnamon stick. Bring to a gentle simmer and cook over low heat, covered, for 15 minutes, stirring occasionally.
2. Add the apple, dates, walnuts, and vinegar and return to a simmer. Cook over medium-high heat, uncovered, for about 5 minutes, or until the liquid has almost evaporated completely, stirring frequently.
3. Adjust the seasoning, adding a little more vinegar if needed. Remove the star anise and cinnamon stick to serve.

Roasted Vegetables with Thyme

These baked Mediterranean vegetables will help get you closer to that ideal of 30 different vegetables a week. Serve warm or cold. These are great for a packed lunch the next day, tossed with salad greens or whole wheat couscous or pasta, and perhaps some halloumi or feta, with nuts and seeds.

250 calories: Serves 4

> 2 bell peppers (any color), seeded and cut into roughly ¾-inch chunks
> 1 sweet potato (about 10 ounces), scrubbed well and cut into roughly ¾-inch chunks
> 2 zucchini, trimmed and cut into roughly ¾-inch chunks
> 1 red onion, cut into 10 thin wedges
> 4 tablespoons extra-virgin olive oil, plus extra for drizzling
> Flaked sea salt and freshly ground black pepper
> 1 tablespoon fresh thyme leaves (2 to 3 sprigs)
> ½ teaspoon crushed red pepper flakes (optional)

1. Preheat the oven to 425°F.
2. Put the peppers, sweet potato, zucchini, and onion in a large bowl and toss with the oil. Season with a large pinch of salt and lots of black pepper. Spread the vegetables in a single layer on a large baking sheet. Roast for 25 minutes, or until the vegetables are softened and lightly browned.
3. Remove from the oven, scatter with the thyme and red pepper flakes, if using, and turn all the vegetables to coat. Return to the oven for another 5 minutes and serve.

Seeded Rainbow Salad

Bitter leaves can help digestion, particularly when eaten at the start of a meal. Here you also get a wide range of anti-inflammatory phytonutrients from the different-colored vegetables.

290 calories: Serves 4

4 ounces mixed salad leaves, watercress, or arugula
2 heads endive (white or red), trimmed and very thinly sliced
4 ounces cherry tomatoes, halved
1 bell pepper (yellow or orange), seeded and thinly sliced
2 cooked beets (not pickled), cut into roughly ¾-inch chunks
2 tablespoons mixed seeds, ideally toasted (see Cook's Tips, page 195), such as pumpkin, sunflower, sesame, and flax

For the nutty dressing:
1 ounce hazelnuts, finely chopped
½ ounce flat-leaf parsley leaves, finely chopped
2 tablespoons apple cider vinegar
6 tablespoons extra-virgin olive oil
Flaked sea salt and freshly ground black pepper

1. Place the salad leaves, endive, tomatoes, and bell pepper in a large serving bowl and toss lightly. Scatter with the beets and seeds.
2. To make the dressing, place the nuts, parsley, vinegar, and oil in a small bowl. Add a pinch of salt and lots of black pepper and whisk together. Drizzle over the salad to serve.

Smashed Chickpeas

This makes a filling and tasty side dish containing lots of lovely sleep-enhancing fiber. Serve with grilled or roasted meats or fish, roasted Mediterranean vegetables, or grilled eggplant.

350 calories: Serves 2

3 tablespoons extra-virgin olive oil, plus extra for drizzling
1 onion, finely chopped
1 garlic clove, crushed
1 (14-ounce) can organic chickpeas
1 teaspoon finely chopped fresh rosemary
½ cup white wine or water (about 1 small wineglass)
Flaked sea salt and freshly ground black pepper

1. Heat the oil in a saucepan, add the onion, and gently fry for about 5 minutes, or until softened, stirring regularly. Add the garlic and cook for a few seconds more.
2. Add the chickpeas and their liquid to the pan, and then add the rosemary and wine or water and bring to a simmer. Cook for 5 minutes, stirring regularly.
3. Remove the pan from the heat and use an immersion blender to blitz the mixture until smooth. Season with salt and pepper to taste. Drizzle with some oil to serve, if you like.

Spinach in Garlic Yogurt

Smooth, softened spinach stirred with creamy, full-fat yogurt makes a great accompaniment to many dishes—fish, vegetables, meat, or curry.

220 calories: Serves 2

> 2 tablespoons extra-virgin olive oil
> 1 small garlic clove, crushed
> 4 ounces young spinach leaves
> ½ cup plus 2 tablespoons full-fat live Greek yogurt or plant-based yogurt
> 2 tablespoons whole milk or plant-based milk
> Flaked sea salt and freshly ground black pepper
> Large pinch of sumac or toasted cumin seeds

1. Heat 1½ tablespoons of the oil in a large saucepan, add the garlic, and gently fry for 20 to 30 seconds, or until softened but not browned, stirring constantly.
2. Add the spinach and cook for about 2 minutes, stirring, until well softened. Transfer to a serving bowl and let cool at room temperature for at least 15 minutes.
3. Add the yogurt and milk to the cooled spinach, then season with a good pinch of salt and plenty of pepper. Stir well.
4. Drizzle with the remaining ½ tablespoon oil and sprinkle with sumac to serve.

COOK'S TIP: Toast the cumin seeds in a dry frying pan over medium heat for 1 to 2 minutes, stirring constantly, to enhance the flavor.

TREATS

Move away from sweetened, sugary, processed foods and snacks as much as you can. They are damaging for your delicate microbiome and, even if you eat healthfully most of the time, you can undo the benefits quite quickly by eating the bad stuff.

The "treats" here have no added sugar; instead, they use whole fruits and the occasional dash of honey or maple syrup to add sweetness.

Fruit is a great source of vitamins and fiber and contains a huge variety of chemicals with antioxidant and anti-inflammatory properties, which are particularly concentrated in and around the skin, so eat as much of the fruit as you can.

Aim to eat a couple of portions of fruit a day—ideally after a meal when they are less likely to spike your blood sugars. Lower-sugar varieties, such as berries and apples and pears, are best. But, as with vegetables, variety is good.

Dorset Apple Cake

A delicious, tangy cake, adapted from the classic Dorset apple cake recipe to give it a gut-friendly twist.

265 calories: Serves 12

2 small to medium eating apples, such as Gala

1 tablespoon fresh lemon juice

1 teaspoon ground cinnamon

3 large eggs

½ cup plus 2 tablespoons butter, at room temperature

7 ounces ground almonds

⅓ cup self-rising whole wheat flour

4 ounces large soft pitted dates, finely chopped

1 teaspoon pure vanilla extract

1 teaspoon baking powder

1 ounce flaked almonds

1. Preheat the oven to 375°F. Grease a 9-inch round spring-form pan and then line it with parchment paper.
2. Core the apples and cut each one into about 12 wedges (keeping the skin on). Place in a medium bowl and toss with the lemon juice and cinnamon.
3. Put the eggs, butter, ground almonds, flour, half the dates, the vanilla, and baking powder in a food processor fitted with the metal blade and blitz until smooth. Remove the blade and stir the rest of the dates into the batter.
4. Spoon the batter into the prepared pan and spread evenly to the sides of the pan. Top with the apple wedges, tucking them fairly close together in concentric circles, and bake for 25 minutes.
5. Remove from the oven, sprinkle with the flaked almonds,

and cook for 12 to 15 minutes longer, or until the cake is cooked through, the apples are tender, and the nuts are golden.

6. Cool in the pan sitting on a wire rack for 30 minutes. Remove the cake from the pan. Cut into slender slices to serve.

Zucchini, Orange, and Apricot Cake

Zinging with orange flavor and remarkably low in sugar, this cake is scrumptious and microbiome friendly. Serve in small squares, like a brownie.

325 calories: Serves 20

4 large eggs

1 cup coconut oil, melted, plus extra for greasing

5 ounces dried apricots, coarsely chopped

2 teaspoons pure vanilla extract

2 small zucchini (about 10 ounces total weight), trimmed and coarsely grated

2 cups self-rising whole wheat flour

5 ounces ground almonds

1 teaspoon mixed spice

1½ teaspoons baking powder

Finely grated zest of 1 large orange

3 balls ginger in syrup, drained and finely chopped

1. Preheat the oven to 350°F. Lightly grease a 9-inch round springform pan and then line it with parchment paper.
2. Put the eggs, melted coconut oil, apricots, and vanilla in a large bowl. Blitz with an immersion blender (or blend in a food processor), until the apricots are very finely chopped and the eggs are pale. Add half the grated zucchini, the flour, ground almonds, mixed spice, and baking powder and blitz again until well blended.
3. Stir in the orange zest, remaining zucchini, and the ginger.
4. Spoon into the pan and spread it evenly so that it fills the pan to the sides. Bake for 30 to 35 minutes, or until risen, golden brown, and firm to the touch.

5. Let the cake cool in the pan for about 10 minutes. Turn it out of the pan and let cool on a wire rack. Cut into small squares to serve.

COOK'S TIP: Mixed spice is available in most supermarkets. It's a mixture of warm, rich spices such as cinnamon, allspice, and nutmeg and resembles pumpkin pie spice, which can be used as an alternative. This cake keeps well for 2 to 3 days wrapped in foil. Extra squares can be frozen.

Banana on Toast with Crunchy Walnuts

If you need an instant filler after a meal, this is easy, healthy comfort food.

325 calories: Serves 1

1 ounce walnut halves, coarsely chopped (a good handful)
1 thin slice seeded sourdough bread
1 tablespoon full-fat cream cheese
1 small banana, sliced
Pinch of ground cinnamon (optional)

1. Put the walnuts in a dry frying pan and toast over medium heat for 2 to 3 minutes, or until hot and lightly browned, stirring regularly.
2. Meanwhile, toast the sourdough bread slice and spread it with the cream cheese.
3. Top with the banana and scatter with the walnuts. Sprinkle with a little ground cinnamon, if you like, to serve.

Nutty Berry Crunch

Healthy berries, nuts, and whole grain fiber. Serve with full-fat crème fraîche or live plain Greek yogurt.

230 calories: Serves 5

About 1 pound frozen summer berries or mixed fresh berries, thawed if frozen (about 2 cups)
2 ounces dried apricots, finely chopped
1 tablespoon honey

For the topping:
2 tablespoons plus 2 teaspoons butter or coconut oil
²/₃ cup whole wheat flour
Heaping ²/₃ cup jumbo steel cut oats
1½ ounces flaked almonds
½ teaspoon ground cinnamon
Flaked sea salt

1. Preheat the oven to 400°F.
2. Put the berries in a shallow 1-quart pie dish. Add 4 to 5 tablespoons water, the apricots, and honey and toss together.
3. To make the topping, using your fingers, rub the butter or coconut oil, flour, oats, almonds, cinnamon, and a pinch of salt between your fingertips in a large bowl until well combined. Sprinkle this mixture over the fruit.
4. Place the pie dish on a baking pan to catch any drips as it bakes. Bake for 30 to 40 minutes, or until the filling is hot and bubbling and the topping is lightly browned. Serve warm or at room temperature.

COOK'S TIP: Try to ensure your berries include strawberries for their natural sweetness.

FERMENTED VEGETABLES

Fun, cheap, and easy to make, home-fermented foods are becoming increasingly popular, valued for their sweet, tangy flavors as well as their rich probiotic qualities.

Fermenting your own vegetables gives you variety, flavor, and far more healthy bugs than almost anything you will find in a supermarket. Experiment with your own variations of different vegetables and spices.

Fill the fridge with jars of colorful vegetables. You may produce more wind as your microbiome adjusts, so eat smaller quantities at first. And use them with caution or avoid them altogether if you have reduced immunity.

The jar(s) need to be clean but not sterilized, and the same goes for your hands. The salt kills most bacteria, other than the acid- and salt-loving bacteria you need for fermenting. Ideally, use organic vegetables.

No calorie counts here—ferments can be considered "free" foods . . .

Purple Pickled Onions

This is one of our favorites—juicy, sweet, and salty pink onion rings bring crunch and flavor to almost any savory meal. Technically, they are fermented rather than pickled in vinegar—which means they taste better, and enhance your microbiome.

¼ small beet, peeled
2 large organic red onions, thinly sliced into rings
3 teaspoons flaked sea salt, such as Maldon
Pinch of coriander seeds
½ teaspoon peppercorns

1. Wash a pint jar with a tight-fitting lid in warm, soapy water. Let it air dry.
2. Grate the beet on a cutting board. Spread a little of the beet over the bottom of a bowl and layer some onions on top of the beets. Continue layering the beet and onions, seasoning with salt as you do.
3. Using your hands (you may need gloves to avoid staining), massage the salt into the onions and beet. Set aside at room temperature for 30 minutes to 2 hours until the juices appear.
4. Stuff the onion mixture firmly into the jar, sprinkling the coriander seeds and peppercorns between the layers as you go. Fill until the jar is full. Add any remaining liquid. Using a blunt object, like the end of a rolling pin or the back of a wooden spoon, press down on the vegetables to release air bubbles and submerge them in the liquid. If not enough fluid is produced to cover the vegetables, add filtered (non-chlorinated) water, 1 tablespoon at a time, to top it up until it reaches ¾ to 1 inch below the top of the jar.

5. Close the lid firmly and keep the onions at room temperature for 5 to 10 days. Release the gases once or twice a day, particularly during the first few days, pressing down to release more bubbles. (This is called "burping.") Finally, refrigerate to slow the fermentation. The onions should last for 2 to 3 months. Remove any bits from the onions that become moldy or blackened. Throw away the onions if they smell bad. The onions should have a slightly sweet, yeasty, and tart smell.

Fennel and Onion

A great way to get two of the best prebiotic vegetables, with all their biome-friendly properties.

2 good-sized fennel bulbs, trimmed and finely sliced
2 medium onions, finely sliced
2 teaspoons flaked sea salt, such as Maldon
1 teaspoon black peppercorns
½ to 1 teaspoon crushed red pepper flakes

1. Wash a pint jar with a tight-fitting lid with warm, soapy water. Let it air dry.
2. Put the fennel and onions in a large bowl, and massage the salt into them with your hands. Set aside at room temperature for 30 minutes to 2 hours for the juices to appear.
3. Stuff the vegetable mixture firmly into the jar, adding any remaining fluid. Using a blunt object, like the end of a rolling pin or the back of a wooden spoon, press down on the vegetables to release air bubbles and submerge them. If there is not enough fluid to cover the vegetables, mix 1¼ cups of filtered or spring water with 1 teaspoon of salt and add to the jar, 1 tablespoon at a time, until it reaches ¾ inch to 1 inch below the top of the jar.
4. Add the peppercorns and red pepper flakes and close the lid firmly. Keep at room temperature for 5 to 10 days. Release the gases once or twice a day, particularly in the first few days, pressing down to release more bubbles. Refrigerate to slow the fermentation. The mixture should last for 2 to 3 months. Remove and discard anything that becomes moldy or blackened. Throw away the fennel and onions if they smell bad. They should have a slightly sweet, yeasty, and tart smell.

APPENDIX: EXERCISES

Resistance Exercises

After the age of 30, most people start to lose muscle; if you are physically inactive, it can be as much as 5% of your muscle mass for each decade that passes. Having more muscle means you look toned, plus it enables you to burn more calories and helps deep sleep. What's not to love? The best way to preserve your muscles is to do weight training or resistance exercises. I have a workout, which I call Fast Strength, and I try to do this most mornings. It works a range of muscles and takes only a few minutes. It is a combination of six exercises that work the top half (push-ups and triceps dips), the legs (squats and lunges), and the abs (crunches and the plank). You should aim to do 30 seconds of each, with 10 seconds' rest in between. Repeat if you feel able.

Push-ups: Lie facedown with the palms of your hands under your shoulders and the balls of your feet touching the ground. Keep your body straight. Lower your body till your elbows form a right angle with the floor and then push up. If you find this too hard, do it with your knees on the ground.

Squats: Stand with your feet apart. Bend from the hips, keeping the weight in your heels. Make sure your back is straight. Keep bending until your legs are at right angles to the floor—imagine you are preparing to sit in a chair. Push back up without bending your back. Squats work the biggest muscles in your body. You can make this harder with weights.

Crunches: Lie on your back with your knees bent, your feet flat on the floor, and your hands by the sides of your head. Curl up your upper body without lifting your lower back off the floor. Make sure your chin is tucked in toward your chest. When your shoulders and upper back are lifted off the floor, curl back down.

Planks: Lie facedown on the floor and then raise yourself onto your forearms and toes so that your body forms a straight line from head to toe. Make sure your midsection doesn't rise or drop. Squeeze your buttocks and hold the position for as long as possible. Remember, it should never cause pain in the lower back.

Lunges: Stand with your back straight and your feet shoulder-width apart. Step forward with one leg, bending both knees to right angles and keeping your upper body straight. Pull back to the starting position and repeat, putting the other leg forward.

Tricep dips: Stand with your back to a bench or chair. Place your palms on the seat behind you, bending your knees to right angles, hips straight. Bend your elbows to right angles to lower your body so that your bottom descends halfway to the floor. Push yourself back up using only your arms.

Vigorous Aerobic Exercise

The standard recommendations are to do at least 150 minutes of moderate aerobic activity (walking, swimming, mowing the lawn) or 75 minutes of vigorous aerobic activity (running, cycling, dancing) a week.

I do a lot of brisk walking (I aim for at least 30 minutes a day) and I also cycle everywhere.

On top of that, I do a very short HIIT workout three times a week. It is ultra-short, but is designed to get your heart rate up. There are lots of proven benefits (for more on these, go to thefast800.com).

I do this workout at home, but it is best, at least to start with, to do it in a supervised setting such as a gym. As with any other form of exercise, it would be wise to discuss with your doctor before starting, particularly if you are on medication.

Michael's HIIT Regimen

My regimen consists of three bursts of 20 seconds, done three times a week on an exercise bike. You should only attempt this once you have built up some fitness. If you are unfit, you should start by doing two bursts of 10 seconds, then slowly build your way up over the course of a few weeks.

1. Get on an exercise bike and do a couple of minutes of gentle cycling, against limited resistance, to warm up. You should just about notice the effort in your thighs.
2. After a couple of minutes, begin pedaling fast, then

swiftly crank up the resistance. The amount of resistance you select will depend on your strength and fitness. It should be high enough for you to feel it after 15 seconds of sprinting.

3. If, after 15 seconds, you can still keep going at the same pace without too much effort, the resistance you've chosen isn't high enough. It mustn't, however, be so high that you grind to a complete halt. It's a matter of experimenting. What you'll find is that as you get fitter, the amount of resistance you can cope with increases. It's not speed but effort you are after.

4. After your first burst of fast sprinting, drop the resistance and do a couple of minutes of gentle pedaling to get your breath back.

5. Then do it twice more.

6. Finish with a couple of minutes of gentle cycling to allow your heart rate and blood pressure to return to normal before stepping off the bike. In total, this takes me less than 10 minutes.

ACKNOWLEDGMENTS

A big thanks to my niece, Emily, for producing the charming and witty drawings for this book. They still make me smile! I would also like to thank all those who shared their sleep stories with me and the numerous academics who shared their knowledge and expertise. And, finally, a big thank you to Aurea and Rebecca, who have provided so much help and support since my first book, *The FastDiet*.

NOTES

1 Prefrontal atrophy, disrupted NREM slow waves and impaired hippocampal-dependent memory in aging. *Nature Neuroscience*, 2013. https://www.ncbi.nlm.nih.gov/pubmed/23354332

2 The effects of partial sleep deprivation on energy balance. *Eur J Clin Nutr*, 2016. https://www.ncbi.nlm.nih.gov/pubmed/27804960

3 Acute sleep restriction increases dietary intake in preschool-age children. *J of Sleep Research*, 2015. https://onlinelibrary.wiley.com/doi/full/10.1111/jsr.12450

4 Associations between short sleep duration and central obesity in women. *Sleep*, 2010. https://www.ncbi.nlm.nih.gov/pubmed/20469801

5 The impact of sleep on female sexual response and behavior. *J of Sexual Medicine*, 2015. https://onlinelibrary.wiley.com/doi/full/10.1111/jsm.12858

6 Sex and sleep: perceptions of sex as a sleep promoting behavior in the general adult population. *Frontiers in Public Health*, 2019. https://www.frontiersin.org/articles/10.3389/fpubh.2019.00033/full

7 Validity, potential clinical utility and comparison of a consumer activity tracker and a research-grade activity tracker in insomnia disorder II: Outside the laboratory. *J of Sleep Research*, 2020. https://www.ncbi.nlm.nih.gov/pubmed/31680327

8 Gender and time for sleep among U.S. adults. *Am Sociol Rev*, 2013. https://www.ncbi.nlm.nih.gov/pmc/articles/PMC4164903

9 A marker for the end of adolescence. *Current Biology*, Vol. 14. https://www.cell.com/current-biology/pdf/S0960-9822(04)00928-5.pdf

10 School start time change, sleep duration and driving accidents in high-school students. *Chest J*, 2019. https://journal.chestnet.org/article/S0012-3692(19)30482-9/fulltext

11 Resetting the late timing of "night owls" has a positive impact on mental health and performance. *Sleep Medicine*, 2019. https://www.sciencedirect.com/science/article/abs/pii/S1389945719301388

12 Entrainment of the human circadian clock to the natural light-dark cycle. *Current Biology*, 2013. https://www.ncbi.nlm.nih.gov/pubmed/23910656

13 In short photoperiods, human sleep is biphasic. *J of Sleep Research*, 1992. https://www.ncbi.nlm.nih.gov/pubmed/10607034

14 Truck drivers should be routinely tested for sleep apnoea. *Guardian*, 2019. https://www.theguardian.com/society/2019/sep/30/truck-drivers-should-be-routinely-tested-for-sleep-apnoea

15 Lifestyle intervention with weight reduction; First-line treatment in mild obstructive sleep apnea. *AJRCCM*, 2019. https://www.atsjournals.org/doi/full/10.1164/rccm.200805-669OC

16 Why lack of sleep is bad for your health. *NHS*, 2018. https://www.nhs.uk/live-well/sleep-and-tiredness/why-lack-of-sleep-is-bad-for-your-health/

17 Sleep patterns and school performance of Korean adolescents assessed using a Korean version of the pediatric daytime sleepiness scale. *Korean J of Pediatrics*, 2011. https://www.ncbi.nlm.nih.gov/pmc/articles/PMC3040363

18 In US, 40% get less than recommended amount of sleep. *Wellbeing*, 2013. https://news.gallup.com/poll/166553/less-recommended-amount-sleep.aspx

19 The sleep habits of an Australian adult population. *Monash University*, 2015. https://www.sleephealthfoundation.org.au/pdfs/sleep-week/SHF%20Sleep%20Survey%20Report_2015_final.pdf

20 Associations of longitudinal sleep trajectories with risky sexual behavior during late adolescence. *Health Psychol*, 2019. https://www.ncbi.nlm.nih.gov/pubmed/31157533

21 Long-term effects of pregnancy and childbirth on sleep satisfaction and duration of first-time and experienced mothers and fathers. *Sleep*, 2019. https://www.ncbi.nlm.nih.gov/pubmed/30649536

22 Menopause and sleep. *Sleep Foundation*. https://www.sleepfoundation.org/articles/menopause-and-sleep

23 Health related quality of life after combined hormone replacement therapy. *BMJ*, 2008. https://www.bmj.com/content/337/bmj.a1190.abstract

24 Physiological correlates of prolonged sleep deprivation in rats. *Science*, 1983. https://science.sciencemag.org/content/221/4606/182

25 The boy who stayed awake for 11 days. *BBC*, 2018. https://www.bbc .com/future/article/20180118-the-boy-who-stayed-awake-for-11-days

26 Night watch in one brain hemisphere during sleep associated with the first-night effect in humans. *Current Biology*, 2016. https://www.cell .com/current-biology/fulltext/S0960-9822(16)30174-9#%20

27 A rare mutation of B1-Adrenergic Receptor affects sleep/wake behaviors. *Neuron*, 2019. https://www.cell.com/neuron/fulltext/S0896-6273(19)30652-X

28 The dangers of doctors driving home. *Guardian*, 2016. https://www .theguardian.com/healthcare-network/2016/jul/26/two-in-five-doctors-fallen-asleep-wheel-night-shift

29 Acute sleep deprivation and culpable motor vehicle crash involvement. *Sleep*, 2018. https://www.ncbi.nlm.nih.gov/pubmed/30239905

30 Daylight savings time and myocardial infarction. *BMJ*, 2014. https:// openheart.bmj.com/content/1/1/e000019

31 Spring forward at your own risk: Daylight saving time and fatal vehicle crashes. *American Economic J*, 2016. https://pubs.aeaweb.org/doi/pdfplus /10.1257/app.20140100

32 Sleepy punishers are harsh punishers: Daylight saving time and legal sentences. *APA PsycNet*, 2017. https://psycnet.apa.org/record/2017-06044-010

33 Effects of initiating moderate alcohol intake on cardiometabolic risk in adults with type 2 diabetes. *Ann Intern Med*, 2015. https://www.ncbi .nlm.nih.gov/pubmed/26458258

34 Beneficial effects of low alcohol exposure, but adverse effects of high alcohol intake on glymphatic function. *Open Access*, 2018. https://www .nature.com/articles/s41598-018-20424-y

35 Before-bedtime passive body heating by warm shower or bath to improve sleep. *Sleep Medicine Reviews*, 2019. https://www.sciencedirect .com/science/article/abs/pii/S1087079218301552?via%3Dihub

36 Can music help you calm down and sleep better? *Sleep Foundation*. https://www.sleepfoundation.org/articles/can-music-help-you-calm-down-and-sleep-better

37 The effects of bedtime writing on difficulty falling asleep: A polysomnographic study comparing to-do lists and completed activity lists. *J Exp Psychol Gen*, 2018. https://www.ncbi.nlm.nih.gov/pmc/articles /PMC5758411/

38 Insomnia: Pharmacologic therapy. *Am Fam Physician*, 2017. https:// www.ncbi.nlm.nih.gov/pubmed/28671376

39 Australian public assessment report for melatonin. 2011. https://www
.tga.gov.au/sites/default/files/auspar-circadin-110118.pdf

40 Circadin 2 mg prolonged-release tablets. https://www.medicines.org
.uk/emc/product/2809/smpc

41 A systematic review of the effect of inhaled essential oils on sleep. *J Alt
and Comp Medicine*, 2014. https://www.liebertpub.com/doi/10.1089/
acm.2013.0311

42 Runner-up: Tim Cook, the technologist. *Time*, 2012. https://poy.time
.com/2012/12/19/runner-up-tim-cook-the-technologist/2

43 This is when successful people wake up. *HuffPost*. https://www.huffpost
.com/entry/this-is-when-successful-people-wake-up_b_596d17a3
e4b0376db8b65a1a

44 The effect of resistance exercise on sleep. *Sleep Medicine Reviews*, 2018. https://
www.sciencedirect.com/science/article/abs/pii/S1087079216301526

45 Effect of breakfast on weight and energy intake: systematic review and
meta-analysis of randomised controlled trials. *BMJ*, 2019. https://www
.bmj.com/content/364/bmj.l42

46 Fiber and saturated fat are associated with sleep arousals and slow
wave sleep. *J of Clinical Sleep Medicine*, 2016. https://jcsm.aasm.org
/ViewAbstract.aspx?pid=30412

47 Effects of diet on sleep quality. *Advances in Nutrition*, 2016. https://
www.ncbi.nlm.nih.gov/pmc/articles/PMC5015038

48 A randomised controlled trial of dietary improvement for adults with
major depression (the 'SMILES' trial). *BMC Medicine*, 2017. https://
bmcmedicine.biomedcentral.com/articles/10.1186/s12916-017-
0791-y

49 Gut microbiome diversity is associated with sleep physiology in humans.
PLoS One, 2019. https://www.ncbi.nlm.nih.gov/pubmed/31589627

50 Effects of a tart cherry juice beverage on the sleep of older adults with
insomnia. *J of Medicinal Food*, 2010. https://www.liebertpub.com/doi
/abs/10.1089/jmf.2009.0096

51 Effect of kiwifruit consumption on sleep quality in adults with sleep
problems. *Asia Pacific J of Clinical Nutrition*, 2011. https://www.ncbi
.nlm.nih.gov/pubmed/21669584

52 Sweet dreams are made of cheese. *Scitable*, 2013. https://www.nature
.com/scitable/blog/mind-read/sweet_dreams_are_made_of

53 Primary prevention of cardiovascular disease with a Mediterranean
diet. *New England J of Medicine*, 2013. https://www.nejm.org/doi/full
/10.1056/NEJMoa1200303

54 Adherence to the Mediterranean diet is associated with better sleep quality in Italian adults. *Nutrients*, 2019. https://www.ncbi.nlm.nih .gov/pmc/articles/PMC6566275/

55 Mediterranean healthy eating, aging, and lifestyles (MEAL) study. *International J of Food Sciences and Nutrition*, 2017. https://www.ncbi .nlm.nih.gov/pubmed/27919168

56 Mediterranean diet pattern and sleep duration and insomnia symptoms in the Multi-Ethnic Study of Atherosclerosis. *Sleep*, 2018. https:// www.ncbi.nlm.nih.gov/pmc/articles/PMC6231522/

57 Fiber and saturated fat are associated with sleep arousals and slow wave sleep. *Journal of Clinical Sleep Medicine*, 2016. https://jcsm.aasm.org/ ViewAbstract.aspx?pid=30412

58 A randomised controlled trial of dietary improvement for adults with major depression (the 'SMILES' trial). *BMC Medicine*, 2017. https:// bmcmedicine.biomedcentral.com/articles/10.1186/s12916-017-0791-y

59 Red meat consumption and mood anxiety disorders. *APA PsycNet*, 2012. https://psycnet.apa.org/record/2012-15003-012

60 A Mediterranean-style dietary intervention supplemented with fish oil improves diet quality and mental health in people with depression (HELFIMED). *Nutritional Neuroscience*, 2019. https://www.ncbi.nlm .nih.gov/pubmed/29215971

61 Scientists bust myth that our bodies have more bacteria than human cells. *Nature*, 2016. https://www.nature.com/news/scientists-bust-myth-that-our-bodies-have-more-bacteria-than-human-cells-1.19136

62 Gut microbiome diversity is associated with sleep physiology in humans. *PLoS One*, 2019. https://www.ncbi.nlm.nih.gov/pubmed/31589627

63 Enhancing influence of intranasal interleukin-6 on slow-wave activity and memory consolidation during sleep. *Faseb J*, 2009. https://www .fasebj.org/doi/10.1096/fj.08-122853

64 Association between maternal fermented food consumption and infant sleep duration. *PLoS One*, 2019. https://www.ncbi.nlm.nih.gov/ pubmed/31584958

65 Effects of probiotics on cognitive reactivity, mood, and sleep quality. *Front Psychiatry*, 2019. https://www.ncbi.nlm.nih.gov/pmc/articles/ PMC6445894

66 Ten-hour time-restricted eating reduces weight, blood pressure, and atherogenic lipids in patients with metabolic syndrome. *ScienceDirect*, 2010. https://www.sciencedirect.com/science/article/pii/S1550413119306114

67 Treatment of chronic insomnia by restriction of time in bed. *Sleep*, 1987. https://www.ncbi.nlm.nih.gov/pubmed/3563247

68 The evidence base of sleep restriction therapy for treating insomnia disorder. *Sleep Medicine Reviews*, 2014. https://www.ncbi.nlm.nih.gov/pubmed/24629826

69 Meta-analysis of the antidepressant effects of acute sleep deprivation. *J of Clinical Psychiatry*, 2017. https://www.ncbi.nlm.nih.gov/pubmed/28937707

70 Exercise to improve sleep in insomnia: exploration of the bidirectional effects. *J of Clinical Sleep Medicine*, 2013. https://www.ncbi.nlm.nih.gov/pmc/articles/PMC3716674

71 Experimental "jet lag" inhibits adult neurogenesis and produces long-term cognitive deficits in female hamsters. *PLoS One*, 2010. https://journals.plos.org/plosone/article?id=10.1371/journal.pone.0015267

72 Melatonin for the prevention and treatment of jet lag. *Cochrane*, 2002. https://www.cochranelibrary.com/cdsr/doi/10.1002/14651858.CD001520/full

73 Using the Argonne diet in jet lag prevention. *Military Medicine*, 2002. https://www.ncbi.nlm.nih.gov/pubmed/12099077

74 Common sleep disorders increase risk of motor vehicle crashes and adverse health outcomes in firefighters. *J of Clinical Sleep Medicine*, 2015. http://jcsm.aasm.org/ViewAbstract.aspx?pid=29921

75 Sleep disorders, health, and safety in police officers. *JAMA*, 2011. https://jamanetwork.com/journals/jama/fullarticle/1104746

76 Optimal shift duration and sequence. *American J of Public Health*, 2007. https://www.ncbi.nlm.nih.gov/pmc/articles/PMC1854972

77 Workplace interventions to promote sleep health and an alert, healthy workforce. *J of Clinical Sleep Medicine*, 2019. https://www.ncbi.nlm.nih.gov/pmc/articles/PMC6457507

78 The impact of sleep timing and bright light exposure on attentional impairment during night work. *J of Biological Rhythms*, 2008. https://pubmed.ncbi.nlm.nih.gov/18663241-the-impact-of-sleep-timing-and-bright-light-exposure-on-attentional-impairment-during-night-work

79 Does modifying the timing of meal intake improve cardiovascular risk factors? *BMJ Open*. https://bmjopen.bmj.com/content/8/3/e020396

80 Shift work, role overload, and the transition to parenthood. *Wiley Online*, 2007. https://onlinelibrary.wiley.com/doi/full/10.1111/j.1741-3737.2006.00349.x

81 Modafinil for excessive sleepiness associated with chronic shift work sleep disorder. *Prim Care Companion J Clin Psych*, 2007. https://www.ncbi.nlm.nih.gov/pmc/articles/PMC1911168

82 Modafinil increases the latency of response in the Hayling Sentence Completion Test in healthy volunteers. *PLoS One*. https://journals.plos.org/plosone/article?id=10.1371/journal.pone.0110639

INDEX

Index

ABOUT THE AUTHORS

Dr. Michael Mosley is a science presenter, journalist, and executive producer. After training to be a doctor at the Royal Free Hospital in London, he spent 25 years at the BBC, where he made numerous science documentaries. Now freelance, he is the author of several bestselling books, *The Fast-Diet, The 8-Week Blood Sugar Diet, The Clever Gut Diet,* and *The Fast800 Diet.* He is married with four children.

Dr. Clare Bailey, wife of Michael Mosley, is a GP who has pioneered a dietary approach to health and reducing blood sugars and diabetes at her surgery in Buckinghamshire. She is the author of *The 8-Week Blood Sugar Diet Cookbook, The Clever Gut Diet Cookbook,* and *The Fast 800 Recipe Book.* @drclarebailey

Justine Pattison, is one of the UK's leading healthy-eating recipe writers. She has published numerous books, makes regular appearances on television, can often be heard on the radio, and contributes to many top magazines, newspapers, and websites. www.justinepattison.com